P9-BJA-493

REGIONAL DEVELOPMENT AND SETTLEMENT POLICY

Titles of related interest

The city in cultural context
J. Agnew *et al.* (eds)

Development and the landowner
R. Goodchild & R. Munton

Development economics and policy
I. Livingstone (ed.)

The development guide
Overseas Development Institute

Development planning
W. A. Lewis

High tech America
A. Markusen *et al.*

Industrial structure and policy in the less-developed countries
C. Kirkpatrick *et al.*

Integration, development and equity
P. Robson

Land policy in planning
N. Lichfield &
H. Darin-Drabkin

Living under apartheid
D. Smith (ed.)

Newly industrialising countries
L. Turner & N. McMullen

Outcast Cape Town
J. Western

Planning, need and scarcity
A. Webb & G. Wistow

Planning in Europe
R. Williams (ed.)

Planning the urban region
P. Self

The price of war: urbanization in socialist Vietnam, 1954–85
N. Thrift & D. Forbes

Production, work, territory
A. Scott & M. Storper (eds)

Regional dynamics
G. Clark *et al.*

Regional economic problems
A. Brown & E. Burrows

Rural housing: competition or choice
M. Dunn *et al.*

Silicon landscapes
P. Hall & A. Markusen (eds)

Technological change, industrial restructuring and regional development
A. Amin & J. Goddard (eds)

Town and country planning in Britain
J. B. Cullingworth

Underdevelopment in Brazil
N. H. Leff

Urban and regional planning
P. Hall

Urban hospital location
L. D. Mayhew

REGIONAL DEVELOPMENT AND SETTLEMENT POLICY

Premises and Prospects

DAVID DEWAR

Urban Problems Research Unit and School of Architecture and Planning, University of Cape Town

ALISON TODES and VANESSA WATSON

Urban Problems Research Unit, University of Cape Town

London
ALLEN & UNWIN
BOSTON SYDNEY

© D. Dewar, A. Todes and V. Watson, 1986
This book is copyright under the Berne Convention.
No reproduction without permission. All rights reserved.

Allen & Unwin (Publishers) Ltd,
40 Museum Street, London WC1A 1LU, UK

Allen & Unwin (Publishers) Ltd,
Park Lane, Hemel Hempstead, Herts HP2 4TE, UK

Allen & Unwin, Inc.,
8 Winchester Place, Winchester, Mass. 01890, USA

Allen & Unwin (Australia) Ltd,
8 Napier Street, North Sydney, NSW 2060, Australia

First published in 1986

British Library Cataloguing in Publication Data

Dewar, David
Regional development and settlement policy:
premises and prospects.
1. Regional planning – Case studies
I. Title II. Todes, Alison III. Watson, Vanessa
361.6'0722 HT391
ISBN 0-04-333023-1

Library of Congress Cataloging-in-Publication Data

Dewar, David, B.A.
Regional development and settlement policy.
Bibliography: p.
Includes index
1. Regional planning – Developing countries.
2. Regional planning. I. Todes, A. II. Watson, Vanessa. III. Title.
HT395.D44D49 1986 338.9'009172'4 86-8063
ISBN 0-04-333023-1

56,449

Set in 10 on 11 point Bembo by Nene Phototypesetters, Northampton
and printed in Great Britain by Billing & Son Ltd, London and Worcester

Preface

This book is a by-product of a research project undertaken by the authors in 1983/4, which had as its focus a critical analysis of regional development strategy, and shifts over time in that strategy, in South Africa. In the course of this work, it became apparent that although settlement policy has long been a central pillar of regional development programmes in many parts of the world and although regional development theory and thinking about settlement policy have both been subject to far-reaching changes in recent years, the two have developed largely independently of each other. This book has been written in an attempt to clarify some of the resulting confusion and to move towards a greater integration of the two areas of understanding.

More specifically, the book provides, first, an overview of recent shifts which have occurred in national and regional development theory and the broader social, economic and political factors which have influenced these shifts. Secondly, it identifies the major policy implications of the various development approaches, with particular emphasis being placed on the rôle of settlement policy. Thirdly, the central differences between policy approaches and the debates surrounding them are identified and discussed. An important theme which emerges in each of these areas is the influence of conditions and ideas in more developed countries on less developed ones. Although exploration of this demands consideration of both, the focus is upon conditions in less developed countries.

David Dewar, Alison Todes and Vanessa Watson

CAMROSE LUTHERAN COLLEGE
LIBRARY

Acknowledgements

The views expressed in this book are those of the authors, who also wish to acknowledge financial support from the Co-operative Scientific Programme of the Council for Scientific and Industrial Research, South Africa. The authors would like to extend thanks to Kathy Forbes who undertook the task of editing and typing the final draft with a great deal of willingness and competence.

Contents

Introduction

The past two decades have been characterized by considerable shifts in thinking in the discipline of regional planning. As more and more has become known about conditions in less developed countries, in particular, it has become apparent that vast numbers of people are locked in conditions of grinding poverty, massive unemployment and complex patterns and processes of inequality.

In the face of this increased understanding, the question of how to alleviate these conditions and how to use scarce and often non-competitive resources to achieve sustainable human and economic development in these areas has become a, if not the, central concern of regional planning internationally. A spate of generalized policy approaches to stimulating positive developmental impulses has been advanced. Many of these, it seems, are tinged with more than a hint of panic as the intractability of the developmental problem in poorer areas becomes increasingly recognized.

Although these approaches are obviously not exclusively, or even primarily, spatial, all, by definition, have spatial implications as policy decisions take spatial form or modify spatial relationships. Almost everywhere, the focus of *spatial* policy is upon settlements. Settlement policy has long been a significant item in the basic basket of policy elements with which national and regional planning have been concerned internationally. There have been very few national or regional strategies which have not, either explicitly or implicitly, had implications for the selective stimulation or creation of settlements.

The focus of this book is on clarifying the implication of different approaches to development for settlement policy. The importance of understanding the relationship between these lies in the fact that in all regions the process of channelling investment in regional space is a continuing one, and, because of the significance placed on settlements historically, it usually occurs around a settlement framework. Thus, on one hand, if the implications are not fully understood, spatial policy may operate independently from, or counter to, the general thrust of the policy approach being employed and may considerably negate the impact of that approach. On the other hand, the characteristics of the spatial structure of a region have implications for determining the appropriateness of particular policy approaches.

The book initially arose as a by-product of a broader research project aimed at evaluating the developmental consequences of policies directed towards modifying urbanization processes in South Africa (Dewar *et al.* 1985). In the course of this project it became apparent that there is a need for a document of this kind: great confusion surrounds the spatial, and particularly the settlement, implications of different approaches to development. Thus, for example, similar developmental rhetoric has often been used, by politicians and planners alike, to support very different settlement policy proposals. Conversely, similar policies or proposals relating to the settlement system have been advanced from fundamentally different developmental perspectives. Even more confusingly, similar settlement policies are advanced to tackle very different types of problems. A number of factors underpinning these confusions can be identified.

Some causes

Formative influences

Perhaps the most important is the fact that a variety of formative influences has, over time, shaped thinking about the rôle of settlements in regional development. These influences, though strongly interrelated, nevertheless show autonomous characteristics in important respects: the result is that the rhetoric surrounding settlement policy seldom reflects any one consistent position. A number of these influences can be identified.

Development and settlement theory Significantly, there has been a tendency for development theory, which focuses on understanding political, social and economic processes of national and regional development, and spatial theory, which focuses on the rôle of spatial structure and settlements in national and regional development, to advance along relatively independent trajectories of theory building. While the distinction is not absolute, the tendency has had two important consequences: on one hand, many of the insights gained from development theory have not been spatially operationalized in practice; on the other, there has been a tendency to grant an artificial degree of autonomy to spatial planning issues.

Further, much of settlement theory is inductively derived: it is strongly tied to, and derives from, specific contexts. Despite this, there has been a pronounced tendency by planners to attempt to apply generalized spatial forms and patterns in very different contexts. The consequence of this has been a blurring of assumptions about the nature of the development problem: similar prescriptions are advanced from different paradigmatic positions and in widely varying contexts.

Interventionist approaches Paralleling this, several theories about how development occurs, and thus should be promoted, have emerged over time, and some of the resulting policy approaches have different implications for the rôle of settlements in stimulating regional development. While the approaches have often been informed by insights gained from development theory, they are not *necessarily* underpinned by an explanatory position: some are reactions more to the perceived failures of previous approaches than theoretically grounded postulations in their own right. Often, however, exponents of one or other of these positions appropriate some rhetoric from different theoretical perspectives to justify their positions, a factor that greatly compounds confusion, since basic assumptions become blurred.

A related aspect of this source of confusion results from the nature and process of theory building. Theoretical conceptualization about development does not occur in a vacuum: it is informed both by the perception of the developmental problem, which tends to shift with changes in economic and political conditions, and by the perspectives and interests of those involved in the process, which also change. Accordingly, broad groupings of theory and perceived solutions tend to occur in a relatively 'epochal' way. At any point in time, different positions tend to be dominant. However, successive positions do not simply subsume previous positions in linear fashion: at any moment a variety of positions, all with vociferous supporters, exists.

Interventionist modes Related to the debate about *what* should be done to promote development is a parallel debate on *how* it should be done: specifically whether development can or should be promoted from the 'top down', that is, by external, exogenous agencies and impulses, or from the 'bottom up', that is, by endogenous, 'grass roots' development.

Again, while there are broad correlations between explanatory and interventionist positions and positions in this debate, the fit is by no means neat. In particular, proponents of very different interventionist positions have used the rhetoric of 'bottom up' development to their own ends.

Political influences

The second major reason for the confusion surrounding settlement policy is the fact that development planning is a highly politicized process. Gore (1984) whose work, *inter alia*, explores the popularity of regional policies amongst governments, argues that regional policies have, despite their theoretical and practical weaknesses, provided certain governments with a useful way of promoting private capital accumulation, while, at the same time, legitimizing the authority of the state by means of the egalitarian rhetoric of these policies. Hence, the rhetoric of regional and national development has often been used to counter developmental political ends, as witness, for example, the case of South Africa, where policies aimed at maintaining political control by a racial minority are increasingly justified through the rhetoric of development planning.

Policy issues

A third cause of confusion is that, over time, different settlement policy issues have become blurred. Settlement policy has, relatively consistently, been moulded by and directed towards four distinct sets of issues:

(a) Settlement policy as an instrument affecting regional economic growth and interregional disparities. This, in turn, has had two areas of focus: an urban–industrial focus, where settlement is seen as an instrument for promoting industrial development, particularly in underdeveloped regions; and a rural focus, where settlement is used as an instrument for promoting and stimulating productive rural development.
(b) Settlement policy as a reaction to urban size. Again, there are two distinct areas of emphasis: a reaction to what is perceived to be the excessive size of larger urban areas; and a reaction to the problem of declining towns or settlements which are considered to be 'too small'.
(c) Settlement policy as a mechanism affecting the level, form and distribution of social and utility services.
(d) A fourth set of issues which has informed settlement policy in many contexts is made up of what may be termed 'extraneous' factors – for example, defence, social control and so on. The

term 'extraneous' is used because these factors, although powerful in their impact in many contexts, are specifically context-related. Accordingly, they are not dealt with in this book.

Clearly, the first three issues are distinct and different. Often, however, similar, and sometimes identical, settlement policy packages have been applied to fundamentally different issues or combinations of issues.

The objectives and structure of this book

Specifically, this book has three primary objectives. First, it seeks to describe and compare the elements of the major approaches to development which have emerged over the past three decades. Secondly, it seeks to bring about a closer tie between interventionist development theory and settlement theory by exploring the implications for settlement policy of different approaches to development. Thirdly, it seeks to identify and clarify the major debates surrounding the use of settlement policies in national and regional development.

The structure which is employed to attempt this is shown in Figure 1. Social, political and economic conditions are considered first, since they have provided the basis for the emergence of new approaches to development. Reactions to problems raised by past approaches are also discussed in each case, as they too have influenced the emergence of new approaches. We then discuss the interventionist approaches to development which have emerged and their implications for regional development and settlement policy. Four aspects of settlement policy are considered: settlement policy as a mechanism for promoting industrial development, settlement policy as a mechanism for promoting rural development, settlement policy as a reaction to urban size, and settlement policy as a mechanism affecting the level and distribution of services. Specifically, therefore, the assumptions underpinning each interventionist approach and the factors affecting shifts are discussed, and the implication of that position for settlement policy analysed, in terms of the four issues identified above. However, regional development and settlement policies, as applied from time to time in different contexts, have not only derived their logic from the broader approach to development being applied. They have also been informed by a number of geographic, economic and sociological theories, the influence of which has persisted over time. The influence of these is discussed

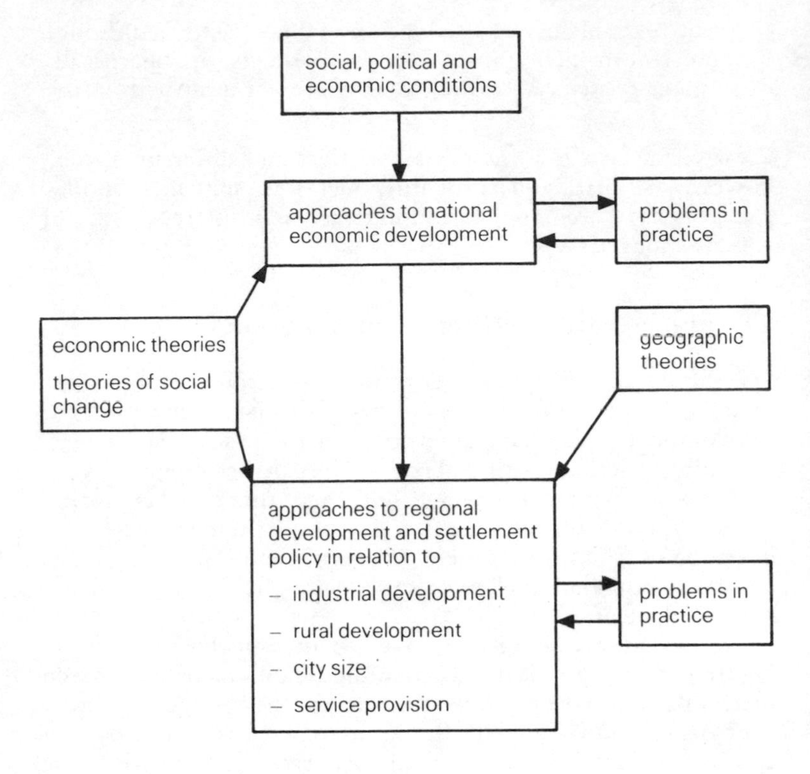

Figure 1 Structure of the book

where appropriate. In the final chapters, major debates relating to the settlement system and its rôle in development are reviewed and some lessons for settlement planning are drawn.

An important theme that is developed throughout this discussion is the influence of conditions in developed countries on national and regional settlement policy in developing countries. On one hand, theories and ideas on the stimulation of regional development have been developed in Europe and the United States of America and transmitted to developing countries, where significantly different conditions prevail. These theories have been advanced as being 'technically neutral': in fact, they are strongly ideologically imbued. On the other hand, developed and developing countries have become increasingly and inextricably intertwined economically. Changing conditions in one group have profound effects on the other. This reciprocal relationship however, is by no means equal: the influence of developed upon developing countries is far greater than the reverse. Despite the emphasis placed upon this relationship, the focus of this text is upon the developing world: conditions in developed countries are discussed only in so far as they throw light upon prevailing attitudes to settlement policy in their developing counterparts.

Similarly, while recognizing that the approaches discussed have been directed towards, and informed by, a variety of political and economic systems, the emphasis is upon approaches articulated in relation to countries and regions which are essentially capitalist. There are two reasons for this. The first is that few developing countries have managed to withdraw totally from the capitalist system. The internal dynamics affecting the transforming space economies and spatial structures of most developing countries reflect, to a greater or lesser degree, capitalist influences. The focus adopted, therefore, arguably allows for a relatively general theoretical examination of the issues involved. Secondly, it is, theoretically at least, easier in socialist than in capitalist contexts, to adjust the pattern of human activities in space, and, provided that resources exist, to take on distributional issues of welfare. This is simply because, in the socialist context, they can be tackled more directly. The focus adopted, therefore, is, in a sense, on the more difficult distributional case.

Finally, throughout the book, we contrast the terms 'developing' and 'developed' to distinguish between poorer and richer nations. We fully recognize the arbitrariness, inadequacy and vagueness of these terms and that more serious attempts at classification may well lead to debates about whether there are in fact three, four (Wolf-Phillips 1979), or even six worlds (O'Connor 1976) contained within these general labels. The task at hand, however, requires a fairly high degree of generalization, and in the text we simply allude to *tendencies* and *emphases* of difference between richer and poorer countries. Consequently, the use of more general terms in this case is less misleading than the use of terms which require more sensitive differentiation.

Having identified what the book seeks to do, it is equally important to state what it does *not* attempt. While it has proved necessary to provide an overview of different perceptions of how national and regional development should be promoted and, broadly, to identify factors which have underpinned shifts in thought, the book does not attempt to provide a systematic review of the history of regional planning. This is done in the first part of Friedmann and Weaver's (1979) work, *Territory and function*. The emphasis here is simply upon the implication of the *mechanics* of different approaches to development for settlement policy. Similarly, no attempt is made to review systematically actual developmental and settlement policies in different countries. In most cases, different interventionist approaches to development only provide the general climate or framework within which actual policy formulation occurs. In practice, policy formulation is profoundly influenced by contextual economic and political conditions. A particularly useful source of country case studies, for example, can be found in Stöhr and Taylor (1981), *Development from above or below?* This work only seeks to uncover the implication of the *inherent* logic of different approaches to development for settlement policy, in order to clarify confusions and to locate major debates about settlement policy formation correctly.

The book also does not attempt to develop some type of regional spatial primer, such as that advanced by E. A. J. Johnson (1970) in *The organization of space in developing countries.* While many of the *issues* raised by Johnson are relevant, we question the degree to which *forms* of spatial organization can be addressed independently of the broad range of contextually specific parameters affecting development and underdevelopment.

By raising and reviewing major differences and debates, we seek merely to provide a framework against which planners, policy makers and other interested parties can analyse particular contexts in order to develop appropriate policy forms.

Finally, we do not seek to develop yet another generalized interventionist approach to development nor a standardized way of thinking about the settlement system. Indeed, the central thrust of the argument is a plea for caution about the use of generalized approaches and a call for more sensitive contextual analyses in order to provide the basis for creative and positive policy design.

PART I

The emergence of spatial planning: 1930s to 1960s

1 *Background to the emergence of spatial planning*

Major forces and influences

The Great Economic Depression of 1929 and 1930, which affected the majority of capitalist countries, generally had a profound and lasting effect on the development of economic and spatial planning. Prior to this, a relatively *laissez-faire* attitude to economic and spatial development had prevailed. The great lesson of the Depression was that capitalist economies were not necessarily self-correcting: they required management and control. Accordingly, a new demand for massive state investment and management to absorb unemployment and to reactivate investment was heard and resulted in political action, particularly in Europe and North America with Roosevelt's 'New Deal' programmes. From this time onwards, the concept of 'top down' planning – that is, state-initiated, planned and implemented action – was increasingly accepted.

However, the effects of the Depression were by no means the same from country to country, or even from region to region within countries. In Britain, for example, the older industrial northern and western areas were particularly hit by the Depression, although their general decline had been occurring since the early 1920s as a result of competition from more mechanized American industries. This resulted in high unemployment in these areas, and accelerated the movement of population to the south-east where newer service and high technology industries – attracted by infrastructure, skills and access to international markets – had been mushrooming. Similar processes of industrial and rural decline were to be noted in certain other developed countries, and regional planning theory as a result became increasingly concerned with the problem of depressed areas – for essentially political reasons. As Alden and Morgan (1974, p. 17) note: 'The movement for regional planning to cope with the problem of depressed areas was essentially concerned with providing jobs, not at all for purposes of economic growth, but to give unemployed men work for social reasons'.

By the early 1940s, however, this essentially reactive orientation of planning had changed. Regional planning had become more orientated towards rational, longer-term management and was increasingly being viewed as an adjunct to national economic planning. Regional planning for depressed areas, in particular, became an important aspect of the Keynesian demand management techniques, which dominated national economic planning at the time. Keynesian policies essentially viewed the control of national demand levels as the key to national economic management. In terms of this logic, therefore, low levels of demand – due to unemployment and low incomes – in the lagging regions were regarded as serious distorting factors.[1] Regional policies were thus implemented in order to promote greater *regional convergence* in employment and incomes, through direct state investment in lagging regions.

The 'long boom' period that followed the economic reconstruction of Europe and North America after World War II introduced conditions that were to have a profound impact on national and regional development policy in both the developed and developing nations of the world. These changes occurred primarily at two scales: national and international.

At a national scale, demand management policies were resulting in a 'stop-go' pattern of economic growth: the pattern of output and demand moved from periods of rapid growth to near standstill and back to rapid growth as governments moved first to encourage demand and then to control it. This, in turn, gave rise to irrationality and waste in long-term investment projects. The increasing size of firms was accompanied by larger, longer-term investments with longer-term repayment periods: firms therefore required longer-term certainty and fuller capacity use. It was in this climate that *indicative planning* arose, to encourage investor confidence and to introduce a greater degree of co-ordination in, and rationalization of, economic decision making (Warren 1972). Indicative planning, which gathered momentum in Europe after the recession of 1958, involved agreement between the three main actors in the economy – the government, the business sector and the trade unions – as to how the economy should function in the long term. It therefore demanded the formulation of medium- to long-term plans – for example, five-year plans – in which the emphasis was upon co-ordinated supply and demand in the economy as a whole and upon 'bringing about structural change in sectoral and regional priorities and specific social expenditures designed to strengthen the civic fabric of capitalism' (Warren 1972, p. 9).

This move towards medium- to long-term plans had significant implications for regional planning. In most of Europe and North America, some sort of regional planning had existed before the advent of full-scale national planning. Without long-term national economic plans, however, regional planning rarely attempted to *modernize* or restructure economically depressed areas or to plan for sustained growth. As indicated, the emphasis – with one or two exceptions, such as Italy – in regional plans until the late 1950s was upon raising the level of employment in depressed regions for political and economic reasons and upon achieving more balanced regional economic growth in an attempt to reduce expected high levels of population growth in metropolitan areas.

With the acceptance of long-term economic planning, regional planning focused increasingly upon the following issues: the creation of an efficient space economy for economic growth; the modernization of depressed areas in line with the demands of national economic restructuring; planning the spatial distribution of population and economic activity; and planning for more strictly political or welfare reasons. Britain in the 1960s provides a good example of this process.

> Nationalization and the loss of colonial markets were seriously weakening the position of some sections of the dominant classes. Together with the need to reorientate the pattern of trade, the growing uncompetitiveness of manufacturing industry, and the impact of the external constraint of economic growth, this led to increasing pressure for a state strategy aimed at the achievement of more rapid economic growth and involving the introduction of an indicative planning system and a reorganization of the state apparatus. These new economic principles were associated with a reconceptualization of the regional problem by the state. This emphasized the importance for national economic expansion of a fuller utilization of labour in areas with high unemployment and low activity rates, particularly in view of the expected emergence of acute labour shortages. By reducing regional inequalities in the distribution of unemployment, the economy would then be able to achieve higher levels of aggregate employment without generating wage inflation through the emergence of tight labour markets. These arguments coincided with an increasing awareness of the fact that the growing concentration of population and employment in the congested areas was beginning to create serious problems, despite the development of the first generation of new towns in the 1950s. Leading sections of industrial capital were therefore arguing for a state strategy of growth, including a more active regional policy that would assist expanding industries to relocate in areas of high unemployment (Dunford *et al.* 1981, p. 399).

An extremely important aspect of the post-war period was a growing concern about national economic development in developing countries. This concern was expressed both internally by the developing countries themselves, and externally by developed nations. Gilbert and Gugler (1982, p. 44) describe these factors:

> The world depression of the 1930s and the effects of World War II led to a spontaneous process of industrial expansion in the larger Third World nations, especially those in Latin America. Protected from imports of manufactured goods from the developed countries, first by the fall in prices of primary exports and therefore a shortage of foreign exchange with which to purchase consumer imports, and later by the shift of industrial production in the developed countries away from export activities towards the war effort, Latin America's industry prospered (Frank 1967). This experience suggested that a new and more successful strategy of development might be embraced. Rather than continuing to rely on imports of manufactured products, Latin American nations could themselves industrialize. . . . Between 1870 and 1930 . . . prices of primary exports from Latin America had fallen relative to the prices of manufactured products. . . By the 1960s no Latin American country had failed to adopt the recommendation [to industrialize]. Import tariffs had been raised, quota restrictions introduced, infrastructure was being erected, and encouragement offered to manufacturing enterprise.
>
> As countries in other parts of the Third World approached and obtained independence, the industrialization/import substitution strategy became conventional wisdom among the new administrations.

The emphasis on the development of developing countries was also prompted by the concerns of developed nations. In part, this concern was fuelled by political factors, particularly the fear of socialist revolution in developing countries (Brookfield 1975) and a desire, once the decolonization process was complete, to establish a basis for collaboration with ex-colonies (Phillips 1977). However, the post-war development initiative in developing countries was also a serious attempt to extend the opening up of the periphery to capitalist relations of production, a factor from which the developed countries in general, and capital in particular, would benefit.

The period prior to World War II had been characterized by structural conditions which tended in general to preserve rather than dissolve pre-capitalist modes of production in developing countries (Phillips 1977). These conditions included the political fear of creating a proletariat, and the subordination of capital to the interests of the nation state in its competition with other nation states. Thus, for example, in the case of Britain, colonies provided cheap raw materials – largely using pre-capitalist methods of production and markets for specifically *British* manufacturing, although this was an obstacle to the expansion and deepening of capitalist relations of production in general. The post-World War II period was an era in which capital was effectively freed from the confines of the nation state. As Brookfield argues, '. . . underlying the last quarter-century of developing the Third World . . . has been the steady and spectacular growth of the "advanced" industrial economies themselves, a growth which has involved a new phase of economic expansionism and integrationalism on a scale far exceeding anything that has gone before.' (Brookfield 1975, p. 25). The persistence of pre-capitalist relations of production in developing rural areas therefore presented a barrier to capital accumulation.

Initially, development strategies focused on *national* economic development – largely to the exclusion of regional policy – which was confined, at this stage, to *ad hoc* projects (Stöhr 1975). From the 1960s, however, policies for regional economic development were put forward, partly to counter the effects of import-substitution policies which had resulted in the spatial concentration of development in the major centres (Gilbert & Gugler 1982), and partly to develop peripheral areas to provide an enlarged economic base for national economic development (Stöhr 1975).

The combined effects of these changes at national and international levels had three significant effects on regional planning. First, the concerns of long-term national and regional economic growth led inevitably to theoretical postulations about how a region grows economically and what constitutes an effective framework for the stimulation of regional economic development. Secondly, the growing concern with developing countries led inevitably to an application of these postulations developed primarily round the experiences of Western Europe and North America – to developing countries. Thirdly, the concern with modernization and the perception of pre-capitalist modes of production as constituting a restriction on capital accumulation led to a 'dualist' conception of developing countries: a conception which identified pre-capitalist relations as 'traditional' and 'backward' and capitalist relations as 'modern'. National or regional development was therefore defined as the transformation of nations or regions from 'backward' to 'modern'. It is this dualist conception of development and underdevelopment which is the hallmark of the modernization paradigm.

In essence, the paradigm, in terms of its implication for the development of underdeveloped countries or regions in developing countries rests upon the following general assertions:

(a) That while 'development' is broadly the process of modernization, or 'an innovation process leading to the structural transformation of social systems' (Friedmann 1972, p. 34), economic growth is none the less a precondition for development. This implies that if economic growth is achieved, the benefits of that growth will automatically be distributed to the advantage of the majority of inhabitants.

(b) That both within and between nations, two economies often exist. On one hand, there is a 'modern' sector focusing upon urban centres of production and exchange, but including the commercial agricultural sector. On the other, there is a 'traditional', largely rural, sector.

(c) That development consists of the incorporation of the 'traditional' into the 'modern' sector – by implication, that the existence of the 'traditional' represents blocks to development, is therefore anti-developmental, and should be overcome.

(d) That the superiority of the 'modern' over the 'traditional' derives from the higher productivity of the former, resulting from higher levels of specialization, division of labour, technology and mechanization.

(e) That the process therefore demands capitalization upon comparative advantage and that in the shift from 'traditional' to 'modern' a critical rôle is played by entrepreneurial skill and innovation. External agents, therefore, are central to the modernization process. Hence, modernization is a 'top down' process, initiated by stronger external agents or sectors.

(f) That urbanization in general and urban centres in particular play a fundamental rôle in economic development because they are centres of innovation and encapsulate and accelerate processes of specialization.

With respect to developing countries, therefore, the modernization paradigm argues that they should – and could – replicate Western experience.

Major growth theories

Two major theories of regional development emerged under the general rubric of the modernization paradigm in the 1950s and 1960s: export-base theory and sector theory. Both were essentially step-downs of theories of national economic development, and, like these national theories, they focused unequivocally upon the stimulation of economic growth, on the assumption that the issue of the distribution of the benefits of growth would be accommodated relatively automatically through the mechanics of the growth process. Before describing each in greater detail, it is necessary to make some broad observations about export-base and sector theories, generally termed 'growth theories'.

First, both were based largely upon historical observations of experiences in the developed world. Broadly, export-base theory drew on the experience of the United States of America, while sector theory more closely reflected the European experience (Stöhr 1974). Implicitly, therefore, both rested upon the assumption that development experience could be replicated.

Secondly, the growth theories were not seen as mutually exclusive. Both were demand-based. One – export-base theory – relates to situations where demand is externally induced: the basic precondition of the theory is an open economy. The other – sector theory – relates to situations in which demand is internally induced nationally or regionally: the precondition is a closed economy. Far from being viewed as mutually exclusive, some theorists (Perloff *et al.* 1960) saw the two together as providing the basis for a broader, more comprehensive theory, though the neglect of the supply side of the equation in both cases was generally seen to be a problem.

Thirdly, in both cases, the shift in theorization from national to regional economic growth was regarded as unproblematic. Both export-base and sector theories were simply regional variants of national theories, and, in practice, regional economic development was seen as largely subordinate to national economic development. Regional spatial planning was primarily concerned with translating (national) development theory into spatial terms.

Finally, the two major growth theories both had significant effects on national and regional development policy internationally, although generally neither was implemented in its pure form. Over time, the impact of sector theory has generally been greater than export-base theory. Indeed, practices from, or the rhetoric of, sector theory, can be found in the development plans of a great many developing countries. Accordingly, greater attention is given to it here.

Export-base theory

Export-base theory derives its theoretical underpinning from the theory of international trade. It argues from the point that regions, like nations, are essentially open: they are tied, to a greater or lesser degree, into a system of regions or nations. In this situation, it assumes that a region or nation develops on the basis of its interregional – or international – comparative advantage, via equal exchange with other regions, and the benefits of growth are seen to spread automatically outwards, sectorally and spatially. While the earliest postulations of the theory, and the experience upon which it was based, suggest that the initial export product will be agricultural or mineral, the theory, at least in principle, does not preclude industrial activity as the 'basic' export activity. The theory assumes that once export-led growth is established, growth is sustained on the basis of a demand-led multiplier process: the development of the export product results in the stimulation of supporting industrial and service activities, and the increase in income which results from increased employment will further stimulate 'residentiary industry' (North 1955). Importantly, therefore, the *success of export-led growth depends on the ability of the region to internalize the multiplier process*, and this in turn depends on the income earned in the export activity, as well as on the lack of extra-regional competing industries which can capitalize on raised effective demand.

Subsequent application of staple theory to export-base theory led to doubts about the universal applicability of export-base theory. Staple theorists, for example Innis (1930) and Baldwin (1956), have argued that the major crop grown in an area – and hence the type of production – would affect the form of development in the area. Hence, 'the nature of the natural resource base, and the institutional form of its development [could] produce radically different conditions for evolution away from an initial export orientation' (Brookfield 1975, p. 95). For example, a region characterized by family firms would be more likely to break from the initial export orientation than a plantation economy, with its characteristically skewed income distribution. This type of observation led to the general criticism that the export-base theory was 'limited in its domain to the quite special case of lands of new settlement, relatively sparse in population and of capitalist economies' (Brookfield 1975, p. 96). Despite these limitations, export-base theory was, and still is, widely used in two primary situations: in situations involving a nationally significant activity with regional implications, and as a regional model involving the *initial* stages of development – that is, the development of exports.

The theory, in principle, has three major implications for the development of the settlement and communications system. First, it demands an outward orientation of the transportation and communication system to ensure close ties with the export market. Secondly, it favours investment in urban centres that are best located in relation to the export market. Thus ports, break of bulk points and so on become important. Thirdly, the emphasis upon comparative advantage implies, with respect to industrialization, capitalization upon economies of scale which, in turn, implies investment in the largest urban centres.

Sector theory

Sector theory derives its theoretical underpinnings from the assumption that economic growth occurs in a number of discrete stages – an approach later formalized and popularized by W. W. Rostow (1960) – and from theories of social change, particularly the work of Dahrendorf (1968).

The concept of stages of growth, based primarily upon interpretations of the experience of western Europe, is founded largely on the empirical observation that rises in per capita income and gross domestic product are associated with shifts in the employment pattern from predominantly primary sectors – usually agriculture – to secondary and subsequently to tertiary sectors. Stages of growth theory explain this phenomenon in terms of the greater degree of specialization and division of labour, and hence efficiency, in each successive shift. It therefore rests heavily on the developmental and innovative capacity of the most advanced sectors (Perloff *et al*. 1960) as well as on the concept of modernization and the dualist characterization of developing economies.

Dahrendorf's theory of social change describes the 'social modernization' effects which accompany the process of economic growth. The process is one of a transformation from a traditional, subsistence-based society where an entrepreneurial consciousness is absent to one where it predominates.

Perhaps the most comprehensive attempt to combine these two strands of theory and to formalize their spatial implications in a closed region can be found in Friedmann's core-periphery theory.

The process is seen to begin with the specialization of agricultural activities within a region and the formation of an intra-regional, urban-centred transport and marketing system to facilitate the distribution of the surplus from agriculture. These urban centres attract existing – rudimentary – secondary and tertiary activities. A hierarchy of urban centres is formed on the basis of the varying levels of diversification of the urban centres and the scale of their market areas. These urban centres, by virtue of their location, become centres of information exchange and, therefore, of innovation. They establish their dominance over their surrounding rural areas and hence attract raw materials and labour from the 'less organized rural to the more organized urban subsystem' (Stöhr 1974, p. 12) and diffuse information from urban to rural areas. The diffusion of information and innovation so engendered allows the urban centre to establish its political dominance over the periphery and this, in turn, results in the infusion of urban values into the rural areas (Stöhr 1974). Although the process of economic transformation is seen as being relatively automatic, once initated, the theory accepts that blocks may occur in the process. The primary rôle of regional policy, therefore, is to initiate the process of development and to remove blockages obstructing the transformation process.

While at a *descriptive* level the theory suggests, like export-base theory, that the initial stage of development should be agricultural, it does not preclude policy intervention at the second level: urban-industrial. Indeed, it gives particular prominence to the 'urban' and to settlement systems and, arguably, provided the primary basis for the predominant use of settlement systems as frameworks for regional development in the 1950s and 1960s.

Note

1 The economic rationale for this was drawn largely from Keynesian economic theory. In essence, this held that the major problem of the depressed economies of advanced capitalist countries was a lack of effective demand. The rôle of the state was, therefore, to boost demand by investment in the productive sectors of the economy. Investment thereby increases employment, either directly or indirectly. Investment in one sector would increase demand for the products of other sectors, either as a consequence of increased employment – for example, in the consumer goods sector – or directly, as a backwards linkage. Demand could also be increased by raising the standard of living of the working class and by improving income distribution.

2 *Implications for settlement policy*

The initiation and diffusion of development

Sector theory has had profound effects on settlement policy throughout the world, although in modified and often distorted form. To understand these and other influences on policy formation it is necessary to examine in greater detail the basic problem posed by the theory to policy makers in the field of regional development, namely, how the processes of development in underdeveloped areas are initiated and diffused. This section examines, first, aspects of economic development theory and settlement theory that have influenced ideas on how development is initiated and how it is subsequently diffused. The specific application of these ideas to settlement policy is then considered in relation to the four concerns that have most commonly informed settlement policy: the desire to stimulate industrial development, the desire to promote rural development, a concern about city size, and the desire to improve levels of service provision.

Initiating processes of development: some economic issues and influences

The question of initiating processes of development, which is clearly a central one for policy makers in the field of regional development, has led over time to subtle but very significant differences of emphasis between developed and developing countries with regard to two interrelated issues.

The first is the stimulation of savings. The interpretation of the problem of underdeveloped areas in both advanced and less developed countries was relatively similar. Theorists identified a vicious cycle of low incomes, low levels of savings, inadequate investment, and thus inadequate capital, and low productivity, which consequently reinforced low incomes (Hansen 1981). The promotion of increased levels of savings, therefore, was identified as being vital in breaking this cycle.

In developed countries, and particularly in western Europe, gross income inequalities were seen as problematic in this regard and policies to reduce income gaps were introduced. Initially, similar policies were attempted in some developing countries but with little success. Following the work of Kuznets (1966), who hypothesized that the more wealthy – and particularly the entrepreneurial class – save more quickly and that inequality is an inevitable and self-correcting characteristic of the early stages of economic growth, policies applied in developing countries either deepened existing inequalities or at least did not challenge them (Brookfield 1975).

The second issue relates to the pattern of investment. For some time – and again based largely on the western European experience – the theory of balanced growth prevailed. This theory held that investment should be diversified over a broad range of sectors and industries. Each industry and sector, it was argued, would generate a demand for the products of other industries and sectors, through increased income levels and direct demands for inputs (Hansen 1981). The theory was soon under attack, primarily for its overemphasis on demand factors and its impracticability: it was argued that it was too costly and too difficult to implement, monitor and administer. By the late 1950s the theory had largely been replaced by the theory of unbalanced growth, a theory widely held to provide the conceptual justification for the 'top down' planning paradigm (Hansen 1981).

Following Hirschman (1958), the theory of unbalanced growth argued that growth strategies should focus investment on relatively few economic sectors, from which growth would 'trickle down' as a result of backwards and forwards linkages. The sectors chosen, therefore, should be the most propulsive – that is, the ones that generate most backwards and forwards linkages. Investment would be concentrated and growth, once initiated, would become self-generating.

Related to these debates were arguments about the allocation of investment between industry and agriculture. Advocates of 'agriculture first' argued that it was necessary to develop agriculture in order to provide markets and funds for urban-industrial growth. Conversely, proponents of an 'industry first' approach argued that the development of industry would result in the absorption of labour from rural areas, allowing agriculture to be modernized. The modernization of agriculture would occur both by the capital-intensification of agricultural technologies and by the diffusion of modern ideas and institutions into rural areas (Gilbert & Gugler 1982). In the 1950s and 1960s, the 'industry first' argument tended to prevail, related as it was to the theory of unbalanced growth.

These influences had a significant effect on settlement policy in both developed and developing countries. In practice, development policies in developing countries tended to focus on urbanization and industrialization, primarily through import substitution policies. This urban bias in policy stemmed largely from five main influences: the emphasis placed upon urban centres in the Rostowian – and Friedmann – conceptions; the importance of urban centres in terms of the theory of unbalanced growth; the experience of urban-based industrialization during the Depression and World War II in more developed parts of developing countries; the interests of the political élite in developing countries (Lipton 1977); and the political desire of leaders of developing countries to 'catch up' with developed countries. Indeed, industrialization generally became the image of development and the symbol of advancement.

Initiating processes of development: some settlement issues and influences

Positions relating to the rôle of settlement in processes of regional and national spatial transformation under the modernization paradigm were informed by two inductively developed strands of theory: rank-size and central-place theory.

Rank-size theory sought to define generalized relationships in the functional size distribution of settlement systems. Building upon early, essentially descriptive, beginnings (Zipf 1949), the theory postulated connections between the level of a country's development and the rank-size distribution of its settlements. In a seminal study of 38 countries in 1961, Berry demonstrated that the relationship was not a simple or static one. Primate distributions, for example, were found both in developed and less developed countries and he found that there were several other variables, such as the country's size and its history of urbanization, which correlated more closely with city-size distribution than with level of development. In subsequent work, El Shakhs, using a different indicator of a country's level of development and a broader index of primacy, found that a relationship did exist between the two, but was not a linear one. He maintained that the relationship follows a normal curve, with 'low primacy at the early and advanced stages of development, and a peak in the middle range' (El Shakhs 1965, p. 47).

Building upon this, Stöhr (1974), in an attempt to relate city-size distribution, interregional disparities and levels of development, postulated that as a country moves from an essentially subsistence to a modern economy, interregional disparities shift from small, to large, to small again. At the same time, the city-size distribution moves from one of many small, unrelated settlements to a situation of primacy in which one large centre dominates many smaller centres. Thereafter the city-size distribution moves to a condition of log-normality in which there is a regular relationship between sizes of settlements and the number in each size rank in the hierarchy, ranging from a small number of large urban centres to a large number of small ones.

The empirical evidence relating to these postulations seems inconclusive. Significantly for policy making, however, and despite this, a log-normal rank-size settlement hierarchy was increasingly regarded as a preferred or more developed, modern pattern – for example, see the proposals for a five-grade hierarchy of settlements in Tanzania (Piöro 1972). The assumption in cases such as these was that if this distribution could be promoted through intervention, economic development would be stimulated.

Central-place theory attempted to explain the size and spacing of central places and the associated patterns and sizes of their trade areas. Central places are distribution points or settlements that provide central goods and services for consumption in the hinterland of the settlement. Their primary spatial organizing principle is efficiency, defined in terms of minimizing consumer movement for goods and services. In terms of the theory, two concepts define the limits within which efficiency can be sought: threshold and range. *Threshold* refers to the minimum amount of support necessary to make the supply of any goods or service viable. *Range* provides a spatial dimension to the concept of threshold. Any good or service has both an upper range, above which it is unable to attract support, and a lower range, below which it cannot be supplied viably. Different goods and services have different thresholds and ranges and, as a general rule, goods and services will be supplied from a minimum number of central places. Space, therefore, is organized around a hierarchy of central places.

The theory was originally developed in the 1930s by a German geographer, Walther Christaller (1966) who focused deductively upon the problem of finding optimizing patterns for three different functions – marketing, transport and administration – and amalgamating these. Although his work was subsequently recognized as being applicable to a limited number of cases only and was greatly extended and sophisticated by theorists such as Lösch (1954), Isard (1960), and in *intra*-urban contexts, by Berry and Garrison (1958), it is Christallerian theory, characterized by an interlocking and nesting hierarchy of different sized settlements and hexagonal trade areas, with each settlement serving a series of six centres of next-lowest order located at the corner points of the hexagon, which has most influenced planning.

Again, existence of a fully developed central-place hierarchy located in accordance with the Christallerian hexagonal geometry came increasingly to be viewed as a preferred state: the assumption was that if such a pattern did not exist, a region was not being optimally served.

The most systematic (and, in terms of regional policy, significant) attempt to link these strands of spatial theory to sector theory, in order to explain how both economic development and the settlement system evolve, was that of Friedmann (1966). The central premiss of his work was that 'for each major period of economic development through which a country passes, there is a corresponding structure of the space economy' (Friedmann 1966, p. 36). The term 'space economy' refers to 'the geographic or spatial pattern of development or the manner in which the economy is manifested spatially' (Fair 1982, p. 8). Central to an understanding of these patterns is a strong 'localizing or polarizing principle in the spatial organization of economic activity' (Lloyd & Dicken 1979, p. 413). By definition, factors of production only come together at a limited number of places. These places form the initial basis of the settlement hierarchy. Once established, the process tends to become cumulative, through processes of specialization, comparative advantage, internal and external scale economies, the concentration of entrepreneurial and political skills and power, and so on. Economic growth is thus, by definition, uneven (Fair 1982).

Over time, this polarization process gives rise to a pattern of core and periphery, with the periphery being dominated by the core. This process of domination occurs in six main ways (Friedmann 1966):

(a) the dominance effect – resource transfers from periphery to core which weaken the periphery;
(b) the information effect – increasing innovation at the core, thereby widening the core–periphery gap;
(c) the psychological effect – creating conditions favourable to continued innovation at the core;
(d) the modernization effect – the transformation of social values, attitudes, behaviour and institutions in the direction of conforming with change;

(e) the linkage effect – innovations breed other innovations by creating new service demands and markets;
(f) the production effect – economies of agglomeration and scale lead to greater growth in the core.

The periphery is not the same everywhere. It 'has two parts: an inner or mobilized periphery, interacting closely with the core, and an outer or unmobilized periphery with weaker links to the core' (Fair 1982, p. 11). This pattern of core and peripheries may occur at various scales: international, between countries; national, between regions; and locally, between urban and rural areas.

This 'dominance–dependence relationship' (Fair 1982) between core and periphery 'is expressed in terms of flows of resources and transactions between them' (Fair 1982, p. 11). Depending on the dominant direction of flow, therefore, the flows can either increase or decrease polarization. Flows from core to periphery are termed *spread* or *trickle-down* effects: if they predominate, polarization is decreased. Flows from periphery to core are termed *backwash* or *trickle-up* effects. If these predominate, polarization increases.

As Fair (1982) correctly observes, the issue of the directional dominance of diffusion flows is one of the essential differences between the modernization and dependency paradigms. The modernization paradigm rests upon the assumption that spread effects will – or can – dominate. The dependency paradigm argues that this is impossible by definition, backwash effects predominate and there is a systematic process of underdevelopment.

The diffusion process upon which sector theory rests does not occur randomly, but hierarchically, through different orders, or levels, of core. Thus:

(a) outward from heartland metropolis to regional hinterlands;
(b) from centres of higher level to centres of lower level in the hierarchy (hierarchical diffusion); and
(c) outwards from urban centres into the surrounding fields (contagious diffusion) (Berry, cited in Fair 1982, p. 12).

Friedmann's conceptualization of the stages of growth consists of four broad periods. Each stage has a characteristic form of space economy, which in turn derives largely from the form of settlement hierarchy. The process commences with a pre-industrial economy[1] in which there is a highly unintegrated pattern of local centres or self-sufficient local areas. Settlements are largely unrelated to each other and perform similar functions. As the stage of transition gets under way, one centre establishes dominance. Since growing manufacture requires localized investment, a 'dualist' structure emerges, comprising a centre and a periphery, the economy of which is either stagnant or declining: the space economy therefore reflects a high degree of primacy.

As countries move through this stage, the gap between core and periphery widens through a process of 'circular and cumulative causation' (Myrdal 1957), by which 'an upward spiralling of growth in the growth centres is contrasted by a downward one (the cycle of poverty) in the stagnant or retrogressive peripheral areas, causing capital and labour to migrate increasingly to the centres of opportunity' (Fair 1982, p. 13).

As the benefits of growth spread – the industrial economy stage – the core–periphery structure changes gradually to a multinuclear structure: smaller centres develop in the periphery and transform it into a series of intermetropolitan peripheries. The space economy therefore becomes more interconnected and integrated and the setttlement functional hierarchy moves closer towards log-normality.

In the post-industrial stage the space economy is completely integrated and is dominated by the urban system: it consists of a series of interconnected city regions and the periphery disappears.

This core-periphery theory, Friedmann (1973) argues, not only applies to economic but to sociocultural and political structures as well. Thus, as this process of transformation occurs, he postulates a political change from a highly centralized to a polycentric system of decision making; a sociocultural shift towards a continuously modernizing surface, characterized by increased inter-group interaction; an economic shift towards increasing decentralization of economic activity and a reduction in inter-regional disparities; and a spatial shift from a high degree of primacy towards an integrated city size hierarchy: a log-normal, rank-size hierarchy. The spatial distribution of the settlement system is strongly informed by the efficiency dictates of central-place theory.

Each of these moves is interrelated and the direction of movement is towards greater equity and economic, social and political integration: counterdirectional moves within any one of the four fields will create tension and conflict within the system.

From a policy perspective, an important aspect of the theory is an attempt to define generic problem regions which lend themselves to relatively standardized prescription. These are:

(a) The core regions or the dominant regions round the major cities or clusters of cities, which are the major generators of change.

(b) The periphery, which has a number of parts:
  (i) Upward transition regions, corresponding with the inner or mobilized periphery – relatively favoured growth areas, usually found around the core regions.
  (ii) Downward transition regions, corresponding with the outer or unmobilized periphery – regions with relatively or absolutely stagnating economies and the source of migrant labour.
  (iii) Resource frontier regions – zones of new settlement.
  (iv) Special problem areas – areas with particular, unique features or problems. The examples Friedmann uses to flesh out this catch-all category include border areas, water resource areas, tourist and military areas, and so on.

Spatially, the resource frontier and special problem areas may occur anywhere within the peripheries but, in terms of the logic of the theory itself, there is clearly a strong probability that they will concentrate in the outer periphery.

In terms of intervention, therefore, sector theory strongly emphasizes the rôle of urban settlements in the development process. Urban centres are central to initiating development impulses, diffusing innovation through space – from the core into the periphery – and internalizing the local multiplier. Further, the theory has strong temporal implications from the perspective of intervention. It is in the second stage of development – the transition stage – that interregional disparities between core and periphery are greatest. This is also the time when the political implications of the disparities are likely to be most strongly felt and hence the time when proposals for intervention are likely to be most favourably viewed by those in power. Significantly, therefore, the classification of the problem areas described above specifically applies to the transitional stage of development.

Diffusing development

The motor of development in sector theory is the diffusion of innovation. 'Innovation' is held to be all advances that further social, economic, political and cultural processes of modernization. In terms of this concept, urban centres and, by implication, the process of urbanization, are vital, *in their own right*, to the diffusion process.

Because of the importance of the concept of innovation-diffusion to sector theory and to regional planning practice in many countries, it is necessary to examine it in greater detail.

Specifically, innovation may be described as 'any new product, technique, organization or idea which is introduced into a social system' (Pedersen 1975, p. 72). From a policy perspective, a distinction needs to be made between two levels of innovation:

(a) *Household innovations*, which are innovations that spread relatively automatically among individuals or private households and that ultimately might be accepted by all, or by large groups within, the total population. Examples are a new consumer good or running water in houses.
(b) *Entrepreneurial innovations*, which are innovations involving considerable risk to the innovator, whether from the private or public sectors. Often this form of innovation is a precondition for the spread and adoption of household innovations. Thus, for example, a precondition for running water in houses is the prior creation of bulk water facilities and there is considerable financial risk in providing these (Berry 1972).

Since, in terms of the modernization paradigm, 'development' involves a process of transformation from 'traditional' to 'modern', the process of development requires flooding the underdeveloped regions with successive waves of innovations and displacing old or outmoded products, techniques, organizations and ideas – politically, economically, socially and culturally.

In a fundamental sense, therefore, the significance of the innovation process derives intrinsically from the dominant paradigm of development. In terms of this paradigm, innovation is the primary motor of development. In an equally fundamental sense, conceptualization of the mechanics of the diffusion process entrenches the essentially 'top down' *nature* of the paradigm.

The pioneering work on the mechanics of the diffusion process was that of Hägerstrand (1953) in the 1950s. On the basis of empirical analysis, he postulated that for the diffusion process to be initiated and to proceed to completion, three essential elements must exist within the social system:

(a) Generators or initiators of information, for example, individuals, households, firms, organizations, settlements, etc.
(b) Receivers or adopters of information, who may be groups or organizations similar to those described above, but obviously operating in a 'less modern' state.
(c) Channels of communication between the generators and receivers. Both information about innovations and the innovations themselves are transmitted through these channels. In essence, the channels are of two types: those that are

dependent upon the entrepreneurial action of establishing a widely based energy supply, particularly radio, television and telephone; and those that depend upon interpersonal contacts. The latter category is considered to be the more important of the two and is affected by the entrepreneurial action of establishing an efficient transportation network, comprising appropriate transportation modes.

Even if these three necessary elements are present, Hägerstrand argued, adoption of innovation will not inevitably occur, for there are barriers to adoption, for example, the friction of distance between generator and receiver and a range of cultural, linguistic, psychological, religious and political barriers. For adoption to occur, these must be overcome. Further, he argued that if the preconditions existed, the time–space pattern of diffusion which is set in motion resembles waves of innovation, losing strength over time and distance away from the generating source.

Significantly, Hägerstrand's initial conception was essentially microsociological. He later suggested that it could be applied to an analysis of spatial diffusion, the key to this lying in the hierarchical ordering of mean information fields (Hermansen 1972b). In the spatial form of the process, the larger cities are seen to play a key rôle as the main generating points. The reason for the emphasis on large cities can again be found in the realm of communications. It is in these agglomerations that the propensity for the processes of discovery, invention and innovation to occur is highest, because of the large concentration of people and the sophisticated communications network, which together facilitate flows of ideas, collusion between innovators, higher probabilities of spontaneous discoveries and so on.

The extent of the wave-like pattern of dispersal around a city defines what can be termed its *communications field*. Clearly, such fields will be more extensive around large cities because of the better communications channels and, in the same way that there is a hierarchy of city sizes, a hierarchy of communications fields can be identified.

Significantly, therefore, from the earliest times, the concept of the diffusion process contained three major premisses which fundamentally affected the 'top down', urban centre-oriented *form* of sector theory:

(a) that innovation impulses develop external to relatively depressed regions;
(b) that innovation diffuses from stronger to weaker elements;
(c) that settlement systems are the key elements in this process.

The central attempt to operationalize the concept of innovation diffusion for interventionist purposes can be found in Berry's theory of the hierarchical diffusion process (Berry 1972). He began from the premise that innovations are essential to regional growth and development and that the city system occupies a central position in innovation and in the diffusion of that innovation. Further, he argued, there is a correlation between the propensity for this to occur and the functional size of cities: thus he tied the diffusion process to the rank-size distribution of cities and towns. The theory of the diffusion process has two main components:

(a) A system of cities arranged in a functional hierarchy, accompanied by processes of information dissemination: hierarchical filtering.
(b) Corresponding areas of urban influence surrounding each of the cities in the system. There is a process of horizontal spreading of innovation outwards into the communications fields, with the spatial extent of the *effects* being proportional to the centre's functional size.

According to the theory, therefore, 'impulses of economic change' (Berry 1972) are transmitted from higher to lower centres in the hierarchy, so that continuing innovation in large centres is critical for the development of the whole system. Areas of economic backwardness are found in the most inaccessible areas, that is, between the least accessible towns in the urban hierarchy. Finally, the growth potential of an area located between any two centres is a function of the intensity of interaction between them. There are four processes which contribute to the hierarchical filtering process:

(a) A *market-searching* process in which an industry exploits opportunities in a sequence of larger to smaller markets and therefore larger to smaller towns: a hierarchical sequence downwards.
(b) A *trickle-down* process in which an industry facing rising labour costs in a large city – because of labour shortages – moves to smaller cities where labour is cheaper. However, because the industry will wish to reduce its trade-off in terms of economies of agglomeration, it will again do so in a hierarchical manner.
(c) An *imitation* process in which entrepreneurs in smaller centres copy the actions of those in larger centres.
(d) A *probability* mechanism in which the probability of adoption depends upon the potential of the adopter learning about the

innovation and this probability decreases with the functional size of a town.

This process is not necessarily rigidly hierarchical. The innovation potential of any town is primarily a function of two factors: its functional rank in the overall hierarchy; and its accessibility relative to other centres which have already adopted. The probability of *household* innovation occurring is governed by these two factors *plus* its position in the communications field surrounding the city – that is, its position relative to the wave-like process of horizontally spread diffusion. Thus, people who are far away from the settlements which are themselves small and low down in the functional hierarchy are often the last to innovate, slowest to change, strongly conservative and so on.

An important aspect of the diffusion process around urban centres is that the process is not infinite. The *hierarchical limit* is represented by the *threshold effect* (Berry 1972): unless an adequate threshold or level of demand exists, the entrepreneurial innovation will be non-viable and will not be introduced – be it in the fields of commerce, health or education. The *horizontal spread limit* is determined by the distance decay effect: it occurs primarily because communications and contact thin out away from the centres.

Application to policy

These various theories, developed in relation to how economic development occurs, how settlement systems emerge and transform economic and social space, and how development impulses diffuse within a nation or region, were to be woven together to form policy instruments for national and regional planners. It is these policy instruments, and in particular settlement policy instruments, that are now examined.

Settlement policy as a mechanism for promoting industrial development

As discussed, sector theory emphasized the significance of the economic core(s) of a country or region in dominating and modernizing successive levels of periphery. To understand how this general emphasis informed settlement policy, however, it is necessary to trace the emergence of another strand of theory: growth pole theory.

The many conceptual contributions to growth pole theory

have been extensively reviewed elsewhere and no attempt is made here to repeat this. However, the work of the following reviewers is recognized: Brookfield (1975), Darwent (1969), Hermansen (1972a & 1972b), Hansen (1981), Lasuén (1972), Parr (1973), Todd (1974) and Thomas (1972). The discussion which follows draws on a number of these works.

Growth pole theory initially emerged against a background of increasing concern in France about the rapid growth of Paris and a perceived overconcentration of activities in the Paris Basin. Significantly, therefore, it emerged primarily as a response to the excessive size and primacy of the metropolis. The response itself was seen to be the development, through concerted national economic planning effort, of a series of *métropoles d'équilibre* or alternative poles of economic activity located some distance from Paris and capable of combating its polarizing effect.

First postulated by Perroux in his seminal article of 1955, the concept was, from the outset, concerned not with geographic but with economic space. There are many conceptions of economic space, argued Perroux, but the one that holds the key to understanding the dynamic processes underlying economic growth and structural change is *economic space as a field of forces*: that is, space defined by poles or foci, emitting and attracting centripetal and centrifugal forces. This repelling and attracting character of each pole defines its relations with other poles. Growth therefore 'does not appear everywhere at once. It appears in points or development poles with varying intensity. It spreads along diverse channels with varying terminal effects to the whole of the economy' (Perroux 1955, p. 310). *Poles de croissance* or development poles are made up of firms, industries or sectors of the economy that have two main characteristics: they have strong internal linkages, through which the centripetal and centrifugal forces flow; and they dominate other firms, industries and sectors and regulate their growth through backwards and forwards linkages.

Perroux's original concept draws on Schumpeter's idea that growth proceeds by the direct and indirect effects of *innovations* (Lasuén 1972). Entrepreneurial innovations, therefore, were perceived to be the prime causal factors behind economic progress. Most innovating activity takes place in large economic units which are able to dominate their environment. This close relationship between scale of operation, dominance and impulses to innovate characterizes the 'dynamic, propulsive firm' of Perroux and the 'leading propulsive industry' of Hansen and Lasuén (Hermansen 1972a).

Schumpeter's theory that development was generated by waves of innovation, and the theory of industrial interdependence and inter-industry linkage form the two cornerstones of Perroux's theory (Hermansen 1972a). In terms of the latter, innovations in one sphere or industry result in clustered responses by entrepreneurs who perceive new opportunities and act upon them. Original innovations have a 'super-multiplier' effect as, for example, highly income-elastic new products replace low income-elastic old ones, and call forth further innovations in supporting industries or industries linked by purchases and sales. Adjustments in linked industries are caused both by linkages themselves – directly and indirectly, through a series of firms – and through the expectations produced by the new product and the perceived impact. By definition, therefore, new industries are those in which innovations occur and which grow faster than other industries and these effects spill over to linked and associated industries (Lasuén 1972).

Further, scale economies play a significant rôle in the growth process. Growth in the leading propulsive industry enables forwards-linked firms to buy their inputs at cheaper unit costs: if the propulsive industry lowers its prices, they consequently tend to expand. Backwards-linked firms, in turn, benefit from economies of scale, as a result of increased production in response to growth in demand from the expanding propulsive industry. Other 'laterally induced' firms whose output depends upon income generated by the propulsive industry and its backwards- and forwards-linked firms also benefit. Capital goods firms benefit from the general climate of expansion but the key to the cycle is the propulsive industry and its pricing strategy. The hallmark of development poles, therefore, is that they consist of large firms, with income-elastic demand curves, strong internal linkages and a position of dominance through their control over backwards and forwards linkage relations.

It is important to recognize from the outset that while the concept of the growth pole is, in its own terms, relatively independent of the dominant growth theories of the time, it is not incompatible with them. In its internal logic it has little overtly in common with export-base theory. However, the implication that poles so developed should be large and dynamic and thus must result in an open-region orientation and a primary reliance on export for the marketing of finished products makes it, of course, compatible with that theory.

Similarly, while there are superficial links with the logic of sector theory, in conceptual terms these are not strong. Thus, for example, while the growth pole is driven by the growth and innovative capacity of the leading sector and rests upon the concept of diffusion of that innovation, both the concept of innovation and of diffusion are inter-industry ones. Growth pole theory says little about the spread to the surrounding region. Further, sector theory specifically rests upon the urban settlement system. As Brookfield (1975) points out, however, Perroux's growth pole concept is only *incidentally* urban: it is not the fact that the poles are urban which makes them dynamic, but their activity and industrial content.

Despite these differences, growth pole theory has, over time, become inextricably linked with sector theory. The first theoretical steps toward tying growth pole theory more specifically into the framework of sector theory can be found in the work of Boudeville and Pottier. Boudeville (1969) argued that there is a spatial equivalent to the concept of polarized economic space – the polarized or nodal region, which is organized about, and interdependent with, a specific centre or set of centres. The implication, therefore, is that the abstract economic forces of polarization take on *physical* dimensions: there is a physical process of polarization. Pottier's (1963) concern was with the spatial pattern of growth. On the basis of empirical observation, he postulated that growth generally follows a linear path or axis, defined by major transportation routes. People, infrastructure and industries locate on the axis and expand along and from it in linear and nodal fashion, with successive rounds of economic growth causing cumulative spatial consolidation along the axis. However, it is Boudeville's theory of polarized regional space which, for Hermansen (1972a), provides the link to theories of location. He argues that Boudeville's theory is not by itself a theory of location which explains where functional growth poles are or where they will occur in geographic space in future. To explain this, growth pole theory must rely on location theories such as central-place theory.

It is chiefly through these interpretations – physical concentration and linear expansion – that the originally analytic, inductive and non-geographical Perrouxian concept has been transformed into a spatial instrument for promoting regional development (Brookfield 1975).

The concept in its new form had immediate and widespread appeal. It was concrete and comprehensible; it offered a tangible framework for investment, which was an important factor in a climate of indicative planning. Most important, it had political promise: constituency politicians could use it as an instrument of hope for depressed areas. Indeed, its very political popularity and its ability to attract a wide range of bedfellows has been an important reason for its practical impotence in many contexts. In this process of transformation from an abstract concept into a policy framework, two interpretative shifts occurred which were to have profound implications for future implementation.

First, the concept of 'innovating activity', defined in general terms in the idea of the growth pole, became synonymous with manufacturing industry and interindustrial economic linkages in particular. Indeed, the complex conceptual definition of a growth pole became identified in practice with groups of manufacturing industries which are 'relatively large, generate significant growth impulses to their environment, [have] a high ability to innovate, and, finally, belong to a fast-growing sector' (Hermansen 1972a, p. 22). In places, the reduction went even further than this: there has been a tendency to ascribe mystical 'growth pole' potentials to almost all, or any, urban places.

Secondly, there was a shift in the type of problem to which the tools were applied. Whereas it was initiated as a reaction to excessive urban concentration – and thus sought to graft new growth onto areas with an existing potential for growth – over time it was increasingly advanced as a tool to tackle the problem of economically depressed regions. It is in this usage that a *de facto* integration of growth pole theory and sector theory occurred.

This integration occurred mainly through three concepts. First, the concept of polarization: economic and spatial polarization were seen as being essentially synonymous and 'pole' became associated with 'urban centre'. Secondly, the concept of innovation and diffusion: the innovation process postulated in growth pole theory unproblematically became associated with the innovation process advanced by sector theory, even though the agents of transmission are entirely different. In the case of growth pole theory, the process is an intra- and inter-*firm* one, and in the case of sector theory it is an inter-*urban* one. Thirdly, the concept of central places, the concept of the spatial organization of the diffusion process in sector theory, provides the framework for the spatial deployment of growth poles. Different levels of central places become associated with different levels of growth poles. A significant by-product of this uneasy fusion has been a tendency to ascribe spatial form to the *abstract*, politico-economic concept of the core–periphery relationship. The abstract economic core becomes the major urban centre(s). This intellectual transference has persisted and has had a profound impact on subsequent settlement policy. In terms of sector theory, urban concentration was regarded as positive in terms of its impact on the surrounding periphery. In terms of later approaches informed more by the dependency paradigm, large urban size has tended to be regarded as negative with respect to its impact on the periphery. In both cases, however, problems of underdeveloped or lagging peripheries have – often entirely erroneously – been interpreted as being related to the size of the core.

The conceptual blurrings have been complicated still further through the process of implementation of policy. At the heart of this difficulty is the fact that growth poles have been used to solve widely differing problems, in a variety of types of lagging regions and at virtually every level of the urban hierarchy. Thus, for example, in developed countries, growth poles have been used to combat perceived excessive concentrations within metropolitan regions; to modernize declining old industrial regions; to adjust or shift relative growth rates in fairly developed regions; to aid *national* economic modernization by facilitating the spatial division of labour within modern industries and/or by facilitating the economically rational location of modern industries; and, to a lesser extent, to modernize regions dominated by pre-capitalist modes of production. In the course of these attempts, a certain confusion has arisen with respect to the content of growth poles. Importantly, this confusion is to some extent present within Perroux's theory itself. In essence, the rationale of growth pole theory is dependent upon attracting the *most propulsive* industries, that is, those that are fast-growing and have many backwards and forwards linkages. These are, by definition, the most difficult to attract, as they are the most entrenched in existing centres. This is the case particularly in developing countries. There is a tendency, both in theory and in practice, to identify propulsive industries with large industries in the most advanced sectors: in practice, it is much more complex than this. While industries in the most advanced sectors may often be fast-growing and may be significant in inducing growth in the economy as a whole, they will not necessarily induce strong linkages or even have broad demand-led multiplier effects.

Increasingly, the most 'modern' sectors of industry are characterized by large companies with the technological capacity to divide their activities in space in order to benefit from a series of 'best fit' locations. One common form of this is the location of head offices and research and development activities in metropolitan areas, semi-skilled functions in larger urban areas, and unskilled or de-skilled functions in peripheral areas.[2] Here, linkages are predominantly intra-industry and there are few ripple effects on the local area (Stöhr & Tödtling 1977).

In many places, however, the emphasis on incentives, which have become a standard part of the growth pole package, and other forms of inducement, have resulted in the attraction, not of 'modern' sectors, but of industries that have limited local growth effects and are sometimes inefficient and unable to stand on their own: most commonly, branch plants, slow-growing 'lame ducks', or industries that are locationally inefficient (Stöhr & Tödtling 1977). Thus in Italy, for example, 'the "propulsive sectors" of the Varoni plan turned out to be those in which the Cassa per Il Mezziogiorno was already undertaking investment and which could hardly be said to be "propulsive" rather than "permissive" in the sense of providing the infrastructure which was a necessary, rather than a sufficient, condition for growth' (Holland 1971, p. 77). In essence, therefore, the quest to attract propulsive economic activities has been subtly transformed into a quest to attract *any* economic activity: the measurement of success has usually been taken to be the number of firms attracted and few poles have achieved self-sustaining growth.

In developing countries, the rhetoric of growth pole theory has been appropriated primarily in relation to three main problems: the modernization of regions which are perceived to be 'backward' in Rostowian terms; the creation of urban centres in association with exploitable natural resources; and the problem of the growth of the largest urban centres. In relation to the first problem, the implantation of urban cores in peripheral areas as a tool of modernization, the cores took primarily three forms: resource or agriculturally related centres, which will be discussed further below; administrative centres – quite often new regional or national capitals; and urban-industrial 'growth poles', or the variant, the 'industrial estate', used primarily in Asia in the 1950s (Lefeber & Datta-Chaudhuri 1971, p. 175).

Generally in the 1950s and 1960s, the urban-industrial growth pole was used as a tool for developing 'backward regions' only where 'overconcentration' in metropolitan areas was perceived to be a problem – a perception which was usually politically based. It was therefore most commonly used in the more industrialized and urbanized developing countries, mainly those in Asia and Latin America.

The record of modernizing, urban-industrial, growth pole strategies in these contexts is extremely poor, primarily because the nature of the problem is very different from that prevailing in the contexts from which the idea has been imported. Many underdeveloped regions in these areas are extremely poorly endowed with competitive resources; they are far from, and poorly connected with, essential national markets; and they do not contain the infrastructural communications, financial, education, health and other back-up facilities necessary for rapid and sustained economic growth. In short, there is little upon which growth can be grafted: in many cases, it would take years of intensive investment in social overhead capital before the strategy would have a chance of success and the attraction of this type and scale of investment *prior* to economic growth occurring is politically difficult. In the face of these difficulties, it is almost impossible to attract genuinely propulsive activities.

In terms of the second problem – the creation of urban centres based on exploitable natural resources – there is often a link between growth poles and export base theory. Indeed, the need for export production – primarily agricultural but sometimes mineral – at a *national* level has led to the emphasis on creating urban centres in association with exploitable resources – that is, mining and marketing towns, and urban centres associated with the transhipment of these resources, such as ports, harbours and railhead towns. The investment in a usually large urban centre is justified in terms of capturing the multiplier process, providing necessary back-up infrastructure for efficient resource exploitation and creating initial linkages for industry. In the long term, therefore, these centres are intended to become alternative core areas within the national space economy, thereby both deflecting migration away from existing cores and setting up a process of modernization within the peripheral region. The Ciudad Guayana project in Venezuela is a widely quoted example of this idea.

Inevitably, the use of export base and growth pole concepts in this way has led to a clear and widening distinction between 'stronger' and 'weaker' regions: the effect has been to exacerbate the existing spatial pattern brought about by colonial production – a pattern of one or a very few growing urban areas – based upon resource exploitation, transhipment or administration, usually located along outwards-orientated transport lines.

Thus growth poles have come to mean all things to all people: the concept has been applied in a wide variety of contexts to widely different problems. This, and the impact of export-base and sector theories on national and regional policy is well illustrated by a purely empirical observation of Stöhr's that

> 'the practice of most national urban development policies appears to combine, in different ways, five basic elements of urban development strategies:

(i) 'New towns or growth poles in peripheral undeveloped regions, usually great distances from the national metropolitan centre.' This generally corresponds spatially with Friedmann's resource frontier regions and ties in economically with export-base theory.
(ii) 'The development of new or intermediate-sized towns as "growth poles" in peripheral lagging regions.' This corresponds with Friedmann's 'depressed regions' in the usually unmobilized periphery.
(iii) 'The development of satellite towns at an intermediate distance (20–100 miles) from the core.' This corresponds with Friedmann's 'upward transition problems regions' in the usually mobilized periphery. This type of action may be designed to reduce congestion in the core area: British and French new towns are examples of this. It is also a reaction to primacy in that it is designed to balance the settlement hierarchy. This latter aspect will be discussed further below.
(iv) 'New urban centres at the immediate fringe of the metropolitan areas'. These correspond with Friedmann's 'core problem areas'.
(v) 'Downtown renewal or in-town new towns' (Stöhr 1974, pp. 23–4).

Clearly, the problems to which these similar actions have been directed are very different. The last three of these policies are reactions to perceived problems of the excessive size of, and concentration within, the core areas. The first two have implications for regional economic growth and interregional disparities. Moreover, the contexts in which these similar instruments have been applied and thus the constraints which they face in their implementation vary widely.

Growth pole theory, innovation-diffusion and sector theory Although the concept of diffusion of innovation in growth pole theory refers primarily to *intra*- or *inter*-firm processes, regional planning theory has tended to assume that similar processes of diffusion of innovation can take place through the urban hierarchy.

With the coupling of the theory of innovation-diffusion and the concept of growth poles, there was a surge of confidence in the field of regional planning. Planners seemed to be in possession not only of the identity of the ingredients of development, but also of the spatial instruments by which the ingredients could be spread to the underdeveloped peripheral regions. Underdevelopment could be defined by a deficiency of innovation acceptance in the backward regions: they were backward because they had not innovated to the same degree as the developed or core regions.

Growth poles could be stimulated in the underdeveloped areas in order to initiate waves of innovation. The direction of innovating processes could be directed through the manipulation of the functional size hierarchy of the settlement system. Principles for the management of a *spatial* ordering of these hierarchical flows could be found in the dictates of central-place theory which defined, in terms of efficiency of marketing, administration and transport, an optimal spatial hierarchy of settlements and thus which indicated both where and in what order the manipulation of the settlement system should occur.

Central to this confidence, therefore, was the belief that innovating cores would have an impact both on their hinterland and on the settlement system. The *conceptual* aspatial work of Hirschman (1958) and Myrdal (1957) indicated that the impact of core on periphery could either be positive, that is, characterized by 'spread' effects, or negative, that is, characterized by 'backwash' effects, but that usually both effects were present.

'Backwash' is a generic term for those processes that accentuate spatial disparities in economic development. These processes may incorporate the following: the selective migration of labour and entrepreneurship from the periphery, which adversely affects the skill mix of the periphery while enhancing that of the growth centre; primary products and savings may be attracted to the growth centre by the possibility of greater returns on capital, such returns stemming from 'economies of scale' with the firm and in the provision of infrastructure; external economies, both of urbanization and localization, which make spatial proximity an asset to linked firms and encourage concentration; and the increase of thresholds which makes viable the progressive attraction of higher-order non-basic activity. 'Spread effects', by comparison, move towards a more equal distribution of development between centre and hinterland. Spread effects include the following: the purchase of food and raw materials from the periphery by the centre; the movement of capital and branch factories to the periphery as the centre becomes congested and factor prices rise; the remittance of wages to the periphery both by long-distance commuters and by migrant workers; absorption by the centre of excess labour and the consequent increase of per capita incomes in the periphery; and the diffusion of growth-orientated attitudes and innovations outwards from the centre.

The extent to which backwash or spread effects predominate around any growth centre will depend largely on the operation of certain contextually related factors: for example, the degree of complementarity which exists between centre and periphery; the relative strength of external economies and diseconomies operating in the growth centre; the adaptability of potential entrepreneurs in the periphery; the nature of human resources movement and the degree of physical integration between centre and periphery. While some theorists (Myrdal 1957) argued that 'backwash' effects would prevail over 'spread', the modernization paradigm in general rested upon the assumption that, in the long term, 'inevitable spread' (Hirschman 1958) would prevail.

While the spatial theory developed provided relatively clear theoretical guidelines for intervention, a number of significant distortions occurred in the process of implementation, particularly in developing countries.

First, as has been shown, a conceptual link was forged between the process of innovation-diffusion and inductively based rank-size theory. Two points need to be made about this linkage. On one hand, the theories were based upon empirical data from the developed world and the results of attempts at empirical validation were at best ambiguous (Stöhr 1974). Nevertheless, the implied association was unproblematically ascribed to developing countries where conditions, particularly in terms of communications, were very different. On the other hand, rank-size theory relates to *functional*, not population, size. Population size was used as a correlate of functional size. This association, too, was unproblematically passed on to developing countries, despite the fact that the correlation between population and functional size in these contexts is undoubtedly far looser. The upshot of these transferences was that an important emphasis in development planning in developing countries came to be placed upon attempting to manipulate the urban size hierarchy in the direction of log-normality, particularly through an emphasis on medium-sized towns. These middle-sized towns usually took the form of new regional capital cities, for example, in Nigeria (Salau 1979) and the Philippines (Prantilla 1979), or of industrial growth centres, for example, Ciudad Guayana in Venezuela (Friedmann 1966). In short, there was a shift away from a concern with *process* and towards an unquestioning acceptance of *form*.

Secondly, and similarly, despite the enormous differences in context between and within, developed and developing countries, there was a naïve acceptance that increased urbanization would automatically result in positive and accelerating processes of innovation-diffusion: the thesis of 'deliberate and accelerated urbanization' (Friedmann 1968) was enthusiastically and largely unproblematically accepted. Again, a detailed concern with process was simplified to the level of a universal acceptance of the form of Friedmann's core-periphery theory.

Thirdly, there was a crudification of the concept of innovation, both with respect to interregional innovations and growth poles. The concept of innovation rests upon a complex and subtle web of communications and information flows: in the theory of growth poles, for example, the primary innovation processes are diffused through the mechanism of the firm and do not necessarily take overt physical form. In translating the theory into policy, however, there has been a simplification of 'communications' to 'transportation', partly due to the crudity of empirically measurable indicators of the diffusion process: the 'necessary precondition' for innovation has, therefore, simply been interpreted as an improvement of the transport system.

An indication of this type of thinking can be found in the flood of regional transport plans, produced in isolation from any broader planning issues, developed in the 1960s in many developing countries (cf. Wilson *et al.* 1966). The implicit assumption underpinning almost all of these was that improved physical infrastructure and physical access was automatically positive or 'modernizing'. Two consequences of this approach have inevitably followed. On one hand, attempts have been made to establish 'innovating poles' where the preconditions for success simply do not exist. On the other, the essentially ambiguous nature of the transportation system in the innovation–diffusion process has been ignored. Transportation has a 'two-edged' effect: it facilitates both backwash and trickle-down effects. In certain situations, however, particularly where the new pole is within the communications and marketing field of an existing and much stronger core, the net effect of opening up the underdeveloped region through improved transport links with the established core has greatly exacerbated backwash: the benefit of most investment in the underdeveloped area is appropriated by the existing core.

Settlement policy as a mechanism for promoting rural development

In practice, the development policies of the 1950s and 1960s, which were based on export-base and sector theories, focused much more upon industrialization than upon rural development.

However, both theories envisaged a rôle to be played by small urban centres, either in encouraging the commercialization of peasant agriculture in underdeveloped areas, or in aiding the modernization of depressed agricultural areas in the advanced capitalist countries. The rôle played by urban centres differs slightly in the two theories.

In terms of the logic of export-base theory, the primary rôle of these towns was twofold: first, to develop marketing functions which would provide outlets for enterprising peasants; and secondly, where appropriate, to form the nucleus of 'nodal' agricultural settlement schemes, the methods of which would diffuse to surrounding areas via a demonstration effect.

In terms of sector theory, urban centres in rural areas would promote modernization by providing markets and marketing facilities, by absorbing people displaced from the land through increased mechanization and by diffusing modernization impulses into the countryside – for example, through improved communications, and research, by facilitating the delivery of essential farming back-up, such as machines, fertilizers and insecticides, and generally by increasing the desire of peasants to enter the cash economy. Spatially, the distribution of settlements would implicitly be governed by the principles of central-place theory.

An examination of rural development practices during this period reveals some variation, both in terms of approaches to rural development and in terms of the rôle played by urban centres. In most developing countries, the thrust of development programmes was towards industrialization, and agricultural development was not seen as a priority. Most agricultural programmes were aimed at modernizing peasant farming or the development of commercial schemes in order to promote export agriculture. The latter usually took the form of estate or plantation schemes or large irrigation projects, sometimes involving river basin development. In most schemes involving peasant farmers, a 'progressive farmer' strategy was used: that is, resources were concentrated on those farmers who already held land or were visibly more successful. Frequently this also meant that rural development was concentrated in those regions of the country that had greater potential in terms of agricultural productivity.

The creation of urban centres was not central to this approach. However, settlements often arose to form the node of irrigation or estate schemes: services were concentrated in these and the settlement formed the point of co-ordination for marketing and farm inputs. The more explicit use of a settlement strategy was to be seen in cases such as the 1959 Pakistan Comilla experiment. This project attempted to 'co-ordinate administrative procedures and practices with the grass roots needs of the rural community' (Lefeber & Datta-Chaudhuri 1971, p. 86). It involved the creation of co-operatives, education and health facilities, mechanization, agricultural extension services and a reorganization of local government. The programme was implemented through the central town of Comilla and the surrounding 400 villages, which were integrated into a settlement hierarchy through the reorganization of administrative systems. Another example is the Indian 'regulated market' system (Johnson 1970). Regulated markets, which took the form of an elected committee to supervise the marketing procedures of certain notified commodities, were set up under British colonial rule in middle-sized towns. According to Johnson, this system increased the demand of farm families for industrial goods, diversified occupations, quickened investment and promoted the integration of town and country.

Less directly, a number of countries established industrial growth centres in the rural areas during this period. The aim of this was not so much to stimulate agriculture directly, but rather to absorb the surplus rural population, in order to encourage agricultural modernization and prevent this rural surplus from moving to the largest urban centres. In a number of countries in Asia, particularly in India, this took on the form of setting up industrial estates outside the largest centres. This programme concentrated on the provision of infrastructure for small-scale, labour-intensive industries in an attempt to encourage people to move off the land.

In developed countries, by contrast, the rural problem was somewhat different in nature: rural areas which had once supported productive peasant farmers were often no longer able to compete with commercialized and mechanized agricultural areas. There had been an exodus of rural population to the industrial cities, and some rural areas had been hard hit by economic depression and war. In France and West Germany, programmes of rural renewal involving the provision of services and improved communications were instigated. In West Germany, the existing hierarchy of urban settlements was used to provide services and industrial job opportunities in order to allow people to remain at least in part-time farming. In most European cases, however, an important solution to the problem of the rural areas was seen to lie not so much in agricultural development as in the attraction of industry and social and physical infrastructure to the rural areas. Such strategies, almost without exception, took on the form of small growth poles.

Settlement policy as a reaction to city size

Export-base and sector theories have somewhat different implications for the issue of city size. Further, policy at this time was also informed by issues that went beyond the logic of either of the two theories.

Implicitly, export-base theory favours the growth of the largest cities, particularly in lagging or underdeveloped areas. In situations where capital and competitive raw materials are in short supply – a hallmark of most underdeveloped regions – economic efficiency, and thus growth, is best served by concentrating investment and capitalizing upon economies of scale. It is only in the relatively rare cases where large sources of unexploited resources exist that the argument may be waived in favour of regional equalization.

Sector theory is somewhat more ambiguous in its attitude towards the growth of the largest centres: different attitudes prevail in different stages of development. Thus, in the early stages of growth, the theory implicitly favours investment in the largest centres: primate centres have the capacity to grow fastest because of market distribution and the higher capacity of these centres to innovate. This position is clearly evident in Hirschman's polarization thesis and has been reinforced by empirical studies correlating the growth of gross national product in developing countries with high degrees of primacy (Mera 1973) and by arguments that in least developed countries, such as those of Africa, the size of total national markets is so small that a fully fledged rank-sized hierarchy of towns could not possibly be supported and that therefore primacy is a logical response to scarcity (Alonso 1972).

In the later stages of growth, however, the theory requires a reduction of primacy and a move towards an integrated rank-size system, as a precondition for the effective diffusion of innovations and the breakdown of peripheries.

Significantly, the logic of the theory says little about size *per se*: it concentrates upon the relative relationship of settlements to each other. The development of a more balanced hierarchy is assumed to occur relatively automatically: overt policy action is only required to speed up the process and to remove blockages inhibiting the operation of the 'natural' forces at work. Implicitly, however, the theory assumes the growing significance of 'diseconomies of scale' with increasing size: the motor behind the changing rank-size structure was identified as being the changing price of factors of production.

In actual policy terms, however, the 1950s and 1960s saw many governments adopting policies to control the growth of the largest cities in both developed and developing countries: the 1950s, in particular, were characterized by a strong 'anti-urban', and particularly 'anti-large city', bias. In the developed world, the reaction against large size was partly the result of strategic considerations brought home by the impact of World War II, but it was also fuelled by predictions of rapid growth of the metropolitan areas, most of which were grossly overestimated,[3] and a growing tendency to equate increased size automatically with increased negative externalities, such as noise and air pollution, increased crime and other social pathologies. This concern over the problem of 'giantism' initiated a continuing search to establish an 'optimum' city size and a continuing debate on the significance and consequences of large city size.[4] Essentially the debate has hinged around the issues of whether and when large cities are desirable and the degree to which both economies and diseconomies of scale are functions of city size or structure, and consequently the degree to which the problems can be managed through the manipulation of urban form, internal decision making, financial and organizational structures, and so on.

Policies to combat the increasing size of large cities have primarily taken four, often interrelated, forms in developed countries: administrative and financial disincentives to new residential and economic growth in secondary, tertiary and quaternary sectors; the relocation of governmental and quasi-governmental functions in alternative centres; the creation of green belts to restrict horizontal city spread; and the creation of new towns to absorb the overspill. These towns are primarily of two kinds: those which are located some distance from the metropolitan core and are thus only loosely connected to it, for example, some of the British new towns, and those towns or 'deconcentration points' which are located on the fringes of the metropolitan area and are usually intended both to be self-sufficient and to capitalize upon the economies of scale generated by the metropolitan region as a whole.

In practice, the creation of the new towns has often been associated with the growth pole concept, with two primary effects. On one hand, the primary concentration has been upon establishing relocated or new *industrial* growth in the new towns.[5] On the other, the problem of reducing the size or growth rate of the core area and the problem of stimulating economic growth in underdeveloped or lagging regions have become fused in the same policy instrument. Often, too, this fusion has been an important factor contributing to the relative lack of success of the new poles: the places that require growth-induction are not capable of attracting the overspill population and functions of existing core areas. Conversely, the places which are most able to attract relatively rapid growth are, by definition, the places which least require assistance: they are neither underdeveloped nor depressed.

Among developing countries, it was primarily the more industrialized countries in Latin America and Asia which tended to encourage economic decentralization, following encouragement by, and experience of, Western planners. The new 'conventional wisdom' of decentralization and growth poles was felt in most of Africa and parts of Asia only in the late 1960s. Significantly, however, apart from a concern about over-urbanization, or urbanization with industrialization, urbanization itself, whether centralized or dispersed, was generally considered to be positive.

The concern with problems of giantism was not the only problem associated with city size: the other side of the coin was represented by a concern over declining towns, particularly in the developed capitalist word. In a number of countries, some small settlements were declining, primarily because of 'epochal' changes brought about by shifts in the dominant form of production. Sometimes the problem of declining small towns was a *consequence* of the application of the growth theories, particularly the export-base theory, where these were intended to induce precisely these epochal changes. The decline of some towns in the Massif Central in France, caused by shifts from small-scale to large-scale capital intensive agriculture, is a case in point. Other types of epochal changes are, for example, a decline in the demand for certain raw materials, such as the collapse of coal-based towns, because of an international swing to the use of oil, and the exhaustion of basic raw materials.

The logic of both theories when applied to this problem suggests that the towns be allowed to decline, but for slightly different reasons. In terms of export-base theory, these areas are by definition 'residual' because of the disappearance of their export base and their historical failure to diversify from that base. In terms of sector theory a town or region can only develop to a point that the balance between population and resources allows. The phenomenon of declining towns, therefore, reflects a sectoral backwardness and is a natural part of the process of maintaining balance. In practice, however, the phenomenon led to political problems and in some cases the state intervened by supporting key services.

Settlement policy as a mechanism affecting service provision

This issue has been important almost since the emergence of regional planning. Its significance stems from a long-held recognition that an important aspect of the quality of life that people experience is the degree to which they have access to social and utility services, for example, clean water, sewage disposal, transportation, education, health, and commercial, cultural and social facilities. Access to services, therefore, is an important aspect of regional inequality. The link between access to services and settlement policy lies in the concepts of range and threshold. Services cannot be provided everywhere in space. They require adequate levels of support or thresholds to be economical, whether those thresholds are measured in terms of finance, levels of use or other criteria. Simultaneously, they have a limited and definable radius of attraction or range; and different services have different ranges: there are certain distances beyond which people cannot travel economically to use a service. By definition, therefore, a certain amount of concentration is required for reasonable levels of service provision. In rural areas, in particular, where population tends to be scattered and thinly spread, settlements which serve surrounding areas become extremely important as holding points for services.

Export-base theory does not address the issue of service provision directly: services result as an *effect* of development. Sector theory, however, provides a more coherent base for service provision. It draws explicit links between development and social modernization, and the delivery of 'modern' services, such as health, education, communications and so on, is central to this. Further, the spatial base of sector theory, that is, central-place theory, provides a framework to guide investment in services. Central-place theories, therefore, have long been a feature of regional planning and have fundamentally influenced investment patterns in many countries: in the Netherlands and Israel, for example, entire settlement systems have been planned on this basis in newly colonized areas. A notable feature in this process has been the relatively uncritical acceptance of the precise geometry of the Christallerian postulation, as opposed to its underlying principles of spatial organization, despite the fact that the key factors such as movement technology, marketing practices and so on have altered radically since the time of Christaller's original theoretical postulation.

In the 1950s and 1960s, *policy* focus on the rôle of settlements in improving levels of service provision was of relatively minor significance and primarily occurred in more developed countries. There were three main areas in which the issue arose:

(a) In declining areas, such as the Massif Central in France and rural Britain. Here central-place theory was used to identify 'key' or 'holding' points as areas where basic services would be guaranteed (Cloke 1983).
(b) As one element of policies to equalize access to certain services regardless of the size of the community, for example, in Canada (Higgens 1972).
(c) As a means of promoting the development of small lagging or 'backward' regions in the absence of larger-scale policies and projects, for example, the Appalachian scheme in the United States of America. Here growth *centres* were used to consolidate investment in services and to provide a limited amount of employment by attracting industry. In essence this fell far short of a growth pole initiative and concentrated mainly upon rationalization by developing the relatively more developed parts of the lagging region.

In developing countries, the issue of service provision has most often been implicitly tied to that of growth poles; the attitude was that the growth pole would *incidentally* provide services for the surrounding rural area as well.

Notes

1 This description of the model is largely based upon Fair's (1982) lucid summary.
2 This is the case in the most advanced sections of the electrical engineering and electronics industry in Britain (Massey 1981). Decentralization on this basis is more common in the advanced capitalist countries, but is usually present in developing countries where decentralization policy is capable of attracting any industry at all.
3 Paris provides an example of this. Paris was predicted to grow to 16 million but has in fact stabilized at 12 million. Projectionists based predictions on a faster rate of population growth than actually occurred and assumed a linearity in the *form* of spatial unevenness. More recently, smaller towns have been growing at the expense of larger ones (Jones 1983).
4 An exercise now widely recognized as being impossible as both costs and benefits are affected by a wide and complex range of factors which are themselves further affected by considerations of physical structure.

5 The French New Towns policy in 1965 represented something of an exception in this regard, for the policy has focused upon the creation of tertiary sector employment to provide the economic basis for growth (Tuppen 1983).

PART II

Changing focus: developments in spatial planning since 1970

3 *Background: major forces and influences*

The period since the late 1960s has been marked by a series of shifts in thinking about the meaning of development, the causes of underdevelopment, and consequently about means for the promotion of development. Specifically, there have been significant paradigmatic shifts in relation to the nature of the development problem and in styles of planning intervention. These shifts in position have been marked by the emergence of a number of new approaches to combating underdevelopment, all of which have some implications for the way in which intervention in the settlement system is viewed as a policy instrument. The spatial implications of some approaches, however, are more far-reaching than others. Further, there is, on some counts, considerable overlap between approaches. For reasons of clarity and brevity, therefore, chapters 3 and 4 provide only a cursory overview of the implications of the approaches for settlement policy. Chapter 5 then focuses upon the current major debates raised by the different approaches, relating to the rôle of settlements in the development process.

The changes in emphasis which took place in development approaches in the post-1960s period were informed both by the experience of past policies as well as by changing economic and political conditions. To understand these changes of emphasis, therefore, it is necessary to view them against a background of interrelated forces and contextual conditions.

Contextual forces and conditions

Changing economic and political conditions

The 'long boom' of the 1960s was followed by recession in the industrialized nations. According to Harrison (1978), one important reason was the rapidity of economic growth itself. This, he argues, resulted in a rapidly expanded demand for labour which, by the late 1960s, was translated into a strong upwards pressure on wages as a result of the increased power of the labour movement. These conditions limited the extent to which governments could regulate economies using conventional, consensus-based indicative planning techniques, and, as profits declined, governments resorted increasingly to short-term demand-management techniques. These demand-management techniques led to short-term economic fluctuations, such as the 1970–71 recession and mini-boom of 1972–73, but by and large, investors lacked confidence, and the policies created highly inflationary conditions. This situation was exacerbated by the collapse of the Bretton-Woods agreement in the early 1970s,[1] and by the quadrupling of oil prices by OPEC countries at the end of 1973. These conditions in the more developed countries had a profound impact on developing countries. This occurred directly through a fall in demand for the products of developing countries, a fall in investment from more wealthy nations and a dramatic rise in the price of oil. It occurred indirectly through the emergence of new strategies attempting to stimulate economic growth in both developed and developing countries. The latter has had a major effect on policy formation in more recent years.

Even before the recession of the 1970s hit the developing nations in any significant way, the wisdom of applying conventional growth theories of development to these contexts was being seriously questioned. By the end of the First Development Decade of the 1960s, it was apparent that while economic growth rates had been relatively high for most of the period, the position of the poorest section of the population in developing countries had worsened in both absolute and relative terms (Sandbrook 1982). According to Dwyer (1977), although the average per capita income of developing countries rose by approximately 50 per cent between 1960 and the mid-1970s, the benefits were very unequally distributed among countries, regions within countries and especially between socio-economic groups. Further, in a number of countries, growth was confined to a 'narrow enclave of modern, organized urban industry' (Streeten 1982b, p. 14), which proved incapable of absorbing sufficient quantities of labour from the subsistence sector. The United Nations reported that in many countries the long-term rate of economic growth actually declined and in others expansion of economic output failed to keep pace with population growth. Nearly a third of the 90 countries surveyed reported declining rates of investment, and food production kept pace with local demand in less than 25 per cent of developing countries (Rondinelli & Ruddle 1976). Further, attempts to promote industrialization led by import-substitution were creating severe problems. Import-substitution processes had been based primarily on the local development of consumer goods industries, but to develop these, capital goods had to be imported. The amount paid out in foreign exchange to obtain these machines led to serious balance of payment problems and imported inflation. In addition, policies to aid industrialization – which was often under the aegis of foreign capital – usually contained biases against agriculture, which employed the majority of the working population in most developing countries: price, tax and exchange rate policies uniformly discriminated against this sector, and these were in part responsible for the slow growth of food production and low incomes in the rural areas (Chenery *et al.* 1974).

A number of political issues which emerged in the 1970s also fuelled the search for new, more appropriate development paths and influenced the articulation of the form of these paths. One such factor was continuing revolution and coups in certain developing countries. The impact of this factor was seen both in the changing emphases of international development agencies and in the rhetoric – and sometimes practice – of leaders of developing countries, who were increasingly forced to respond to the reality of continuing mass poverty. A second factor was the growth of demands by leaders of developing countries for a New International Economic Order, a demand fuelled by the success of OPEC. This demand has formed the central focus of most United Nations Conference on Trade and Development (UNCTAD) meetings in the 1970s and 1980s. According to Streeten, the root of the call for a New International Economic Order lies in the 'dissatisfaction with the old order which, it is felt, contains systematic biases perpetuating inequalities in power, wealth and incomes and impeding the development efforts of the developing countries' (Streeten 1982a, pp. 3–4). For leaders of developing countries, development aid had been insufficient in quantity and of poor quality. Moreover, it had been declining since 1968. Aid was usually conditional, and enabled donors to influence the course of recipients' economic and political affairs. Further, trading practices discriminated against manufacturers in developing countries, and exchange rate policies promoted the interests of developed countries.

Advocates of a New International Economic Order therefore demanded *international* policies that would remove biases in favour of developed countries and would promote the economic growth of developing nations. Paradoxically, the main response to this increased pressure has been a tendency for planners and economists from *developed* countries to advocate 'self-reliant', 'autonomous' development for underdeveloped countries and regions.

Nonetheless, poor economic growth rates were to continue in both developed and developing countries throughout the 1970s and into the 1980s. Export revenues declined as world trade dropped off. Tight money supply in the developed countries severely affected poorer countries, forcing them to borrow heavily on international credit markets (Cleveland & Brittain 1978/9). By the end of the 1970s, international banks were becoming increasingly concerned about debt recovery from these nations. This issue peaked in the early 1980s when the inability of Mexico to repay enormous foreign loans threatened the collapse of the international banking system and sent shock waves through the financial currency and stock exchanges of the developed countries.

A significant exception to this pattern has been the economic performance of a group of developing countries called the Newly Industrializing Countries, or NICs. One response to the economic slowdown in developed countries has been for certain multinationals to reduce their labour costs by relocating sections of their production process to labour surplus countries such as Taiwan, Hong Kong, Singapore and South Korea. Foreign investment, together with economic policies which avoided interference in the operation of 'free market forces', resulted in high economic growth rates which have only recently shown signs of slowing down. While changing economic conditions in the early 1970s were reflected in many quarters in a shift in development concerns – initially to the distribution of benefits of growth, and then to the form of growth and the question of poverty – the performance of the NICs provided the stimulus for promoting an 'accelerated growth' approach to development in a number of countries.

The influence of international lending agencies

In practice, international agencies, and particularly lending agencies,[2] have exerted a major influence on the generation of development strategies and upon the selective implementation of strategies, mainly as a result of the shortage of finance in a number of developing countries. The policy of international lending agencies at any point in time reflects the varying influences discussed above, as well as the interests of the lending agency itself. Two tendencies, in particular, are noteworthy. The first is the fact that these agencies have influenced not only the *form* of programmes and projects, but actual priorities and development paths. The usual *modus operandi* is for the lending agency to send a team of 'experts' into the recipient country to 'analyse' the nature of the problem and to 'suggest' strategies. Inevitably, the 'analysis' is biased by the strengths and preferences of the institution itself. Further, 'the suggestion' of policies and programmes is usually more deterministic than the term implies: the award of loans is often directly or indirectly conditional upon the suggestions being accepted. It is not surprising, therefore, that two widely used approaches to development – redistribution through growth and basic needs – have not only been popularized by the International Labour Organization, or ILO, but actually formulated by its economists. Similarly, in the 1980s, the World Bank, in its post-McNamara period, has been largely responsible for the promotion of the accelerated growth approach. International agencies, therefore, have become an important motor in the *process* of generating development approaches. The second tendency, prevalent in the 1980s, has been an increased stress placed by all agencies upon 'efficiency' in loan use, 'productivity' of investments and, in particular, upon the ability of recipient countries to repay loans. One inevitable consequence of this has been the inability of the least developed countries and regions – the areas which most require assistance and which are least able to generate internal investment – to gain access to finance.

The relative failure of regional policies

The economic crisis in the developed countries reinforced an emerging critique of existing regional policy based on the modernization paradigm. In particular, the economic crisis tended to exacerbate existing forms of uneven development and it exaggerated and brought to the fore newly emerging forms of unevenness. Where it appeared that some sort of regional convergence had been occurring, divergence re-emerged. In Britain, for example, the

> emergence of a new international division of labour and restructuring in the textile industry together with the internationalisation of steel production, the development of new steel complexes in maritime industrial zones, and the reduction in demand, exacerbated by the particularly marked collapse of steel-using industries in the U.K., are being translated into extremely large job losses in declining industrial regions (Dunford *et al.* 1981, p. 383).

In countries with strong initial disparities, particularly, regional unevenness was exacerbated by the crisis. In some developed

CAMROSE LUTHERAN COLLEGE
LIBRARY

countries, a problem of the declining inner city has emerged. In some cases, such as the United Kingdom, the decline is due to the 'modernization' or restructuring of industries previously located in inner city areas. In other cases, the rapid evolution of the advanced industrial economy has been associated with spatial shifts in the location of the leading sectors and a concomitant active decline of other areas (Perry & Watkins 1981).

Politically, continuing problems in traditionally depressed regions, fuelled in part by the failure of regional development programmes to reduce uneven development, resulted, in certain countries, in the emergence of regionalist movements. As Dunford *et al.* note, this 'unevenness often leads those living in the most depressed areas to see their problems as local ones, that could be alleviated by exerting political pressure on the state to secure policies aimed at meeting their specific needs' (Dunford *et al.* 1981, p. 377). In some cases, these depressed regions have been ethnically defined, for example, the United Kingdom's Celtic fringe, and this has led to the emergence, or re-emergence, of 'regional nationalism' (Edel *et al.* 1978). The most significant impact of these movements has been an increasing call for 'autonomy' from the central state. For some planners who have been impressed by these movements, 'autonomy' and regionalism – regional self-determination – are seen as a central goal of development.

These conditions led to a re-examination of many of the fundamental concepts of the growth theories, particularly the assumption of inevitable regional economic convergence, and a linear, predictable path of development. By implication, therefore, many of the assumptions underlying regional policies were questioned. This process was strongly reinforced by an increasing number of empirical studies in both developed and developing countries which showed that regional policies had yielded disappointing results, even prior to the crisis. A number of highly consistent conclusions emerged, particularly, though not exclusively, in developing contexts.

First, there was no clear evidence that policies had resulted in regional economic convergence. Where some convergence had occurred at one level, usually the interregional level, divergence had usually occurred at another, generally the intra-regional or interpersonal level (Stöhr & Tödtling 1978). Policies had been most successful where initial regional inequalities were small, but they had had little positive effect where disparities were strong. In many cases, initially perceived convergence proved more apparent than real: although convergence was statistically measurable, this was due mainly to distortions, such as the reduction of employment in 'core' areas as a result of economic contraction or the narrow definition of the problem. For example, statistical measures specified employment and income, but not the skill levels, of employment created (Massey 1978).

Many countries had relied on growth centre strategies in an attempt to reduce regional inequity, but the results of this policy were disappointing. Empirical analyses revealed that growth pole policies tended to attract mainly slow-growing industries or 'lame ducks', firms from elsewhere in the periphery, and capital-intensive branch plants or subsidiaries characterized by extra-regional control or ownership. Branch plants in particular have tended to have more detrimental than positive effects on the welfare of the recipient location. Because of their large share of routine production processes and the lack of their own administrative and research activities, branch plants very often create relatively low-skill and thus low-paid jobs. Smaller branch plants show more instability with respect to macro-economic fluctuations because in times of crisis their headquarters tend to reduce employment in the peripheral branches first. Finally, branch plants tend to have linkages over larger distances, and therefore make fewer regional purchases, have a smaller multiplier effect, and contribute to leakage out of the region (Stöhr & Tödtling 1978).

Settlement policies, therefore, have often not truly aided convergence, but they have produced new forms of uneven development (Massey 1978). In addition, it was increasingly agreed that it was virtually impossible to distinguish between the effects of regional policy *per se* and the normal effects of the 'market mechanism'. For example, many argue that regional policies in more developed countries have, on the whole, tended to facilitate processes which would, over time, have occurred without the help of those policies.

Secondly, studies of the impact of settlement policies, particularly growth pole policies, and policies that rely upon the implantation of core regions in peripheral areas, indicated that the impact had often been negative. 'Spread' effects were usually smaller than expected, or less than 'backwash' effects (Stöhr & Tödtling 1978). These results were corroborated in both developed and developing countries; for example, Moseley's (1973a) regression and trend surface analysis of the impact of the town of Rennes on its surrounding growth space led him to conclude that the spatial extent of the 'upward transitional area' within which spread effects occurred was confined to a radius of some 20 to 30 kilometres from the centres, or the extent of the commuting hinterland. Beyond this, the more remote rural areas were characterized by 'population decline, with pronounced nett out-migration and decline in the active work force; reduction in the number of small farms and a marked decline in agricultural employment; a declining number of food shops and of building trade establishments, with little or no injection of new industrial activity; relatively low levels of car ownership and of domestic amenities' (Moseley 1973a, p. 73). Robinson and Salih's (1971) trend surface analysis of levels of development around the city of Kuala Lumpur indicated a similar upward transitional zone within a radius of ten miles or so of the city. Beyond this, the area was in downward transition. Finally, Gilbert, in his study of Mendellín, Colombia, showed that 'the development "scores" in the 25 kilometre band immediately contiguous to Mendellín were far higher than those elsewhere and that the "scores" in the areas outside this band fell consistently according to their distance from Mendellín' (Gilbert 1975, p. 329). Gilbert (1975, p. 331) drew two important conclusions from his reviews of growth pole impact studies:

> The first is that social services and infrastructural improvements do not diffuse from growth centres beyond a certain limited area, whether the region be located in a rich or less developed nation. The second is that, either as a result of weak economic "spread" effects and/or substantial "backwash" effects, the regions beyond the immediate vicinity of the growth centre received little in the way of positive economic benefit.

Another study by Moseley (1973b) took further an examination of the mechanisms by which spread effects are supposed to occur. Focusing more particularly on the economic impact of a growth pole, he undertook an empirical analysis of the processes of labour recruitment, the purchase by industry of materials and services in the hinterland and the expenditure of income generated by industrial activity. His conclusions supported his previous findings and went some way towards explaining them. First, commuting hinterlands had remained essentially static over time and migration to the growth pole had been mostly short distance. Extra-industrial activity had been generated by regional purchasing, by the use of local subcontractors and by the establishment of small branch factories; however, 'the scale of such impact is small because most of the larger firms in the centres have been unaffected. Much the greater proportion of "spin-off"' has been outside the region' (Moseley 1973b, p. 92). Further, within the region, 'the industrial activity generated in other towns appears to relate directly to their importance as industrial centres and inversely to their distance away' (Moseley 1973b, p. 92). Similarly, the generation of service activity had been channelled primarily to the larger centres, and villages and small towns had largely been unaffected.

Thirdly, it was becoming evident that, in many cases, the development of growth poles was re-creating, at a regional level, precisely the same problems of peripheralization that they had been created to combat at a national level: by creaming off the strongest human, natural and financial resources from the region, they were systematically deepening underdevelopment in the most marginal subregions (Stöhr & Tödtling 1978).

Finally, it was realized increasingly that the reduction in national economic growth rates in the 1970s would make regional policies even less effective than they had been, as the policies were dependent upon redistributing national economic growth spatially and as less funding would be available for investment.

The primary effect of these assessments of regional policy was a widening debate about the most appropriate form of regional policy. Increasingly in developing countries, regional and settlement issues were tied to new development strategies which were intended to be more strongly rooted in the existing resources and conditions of their local contexts, and which were usually perceived to provide more direct benefits to the poor. In the developed countries, while an emerging lobby has argued for more regionally centred development strategies, the dominant tendency in practice has been simply to modify previous policies.

Intellectual forces and conditions

The impact of dependency theorists

During the early 1970s, a new body of explanatory development theory, which was to have a profound impact on national and regional planning in general and, to a lesser degree, on settlement policy, was being popularized. This theoretical perspective, commonly subsumed under the title of the dependency paradigm, had its foundations in economic *structuralism* and was, at least in terms of some of its central concepts (Berger 1974), strongly informed by Marxist conceptualizations.[3]

Economic structuralist interpretations, in general, argued that problems of underdevelopment were not simply a result of existing blocks to development, but derived from the economic structural relations between underdeveloped and developed countries and regions under capitalism. The dependency paradigm, which initially was developed largely through the work of economists engaged in the United Nations Economic Commission for Latin America and popularized by A. G. Frank (1971), built upon this position and, *inter alia*, mounted a broad-based attack on the very foundations of the modernization paradigm, both conceptually and from the point of view of intervention.

At a conceptual level, it challenged some of the fundamental assumptions upon which the modernization paradigm was based: the assumption of a linear path of development, the assumption of a dual economy, and the assumption of the relatively automatic diffusion of developmental benefits downwards and outwards into the periphery. These assumptions, it was argued, are empirically invalid and conceptually misleading.

The assumption of a linear path of development The assumption that the path of economic development of developing countries would automatically follow that of the developed industrialized nations, it was argued, is not borne out empirically. In fact, the development path of developing countries is different in a number of important respects. In particular, it is characterized by a form of *dependent development.* The nations of the world are not independent entities: they are tied into a 'world-wide interdependent system' (Brookfield 1975, p. 24) in which 'the economies of one group of countries are conditioned by the development and expansion of others' (Dos Santos 1973, p. 76). The dependence of developing countries is thus a *structural* economic one. It is characterized by unequal exchange and technological dependence; also by financial transfers from developing to developed countries in the form of loan and interest repayments; and in the transfer of profits by foreign firms. Further, these foreign firms are usually capital-intensive and consequently employ few people or else employ them in low-skill jobs. This limits the development of a local mass market and reinforces the demand for sophisticated consumer goods with a high import content. This pattern, it is argued, bears little resemblance to the history of developed countries. Finally, while there are variations in the emphasis and form of the argument, it is held in essence that while it is primarily economic structures that determine dependence in some cases, these are underpinned by the inability of a modernizing bourgeoisie to emerge. This, in turn, prevents the emergence of an autonomous form of capitalism which might in some way follow the path of developed countries.

The assumption of a dual economy The conception of developing countries being made up of two relatively independent economies, one 'traditional' and one 'modern', has been attacked on empirical grounds. Dependency theorists argue that by implicitly characterizing developing countries as 'traditional', pre-colonial or suffering from original underdevelopment, the modernization paradigm ignores many cases in history of relative abundance. Historically, in these cases, the self-sufficiency of the 'traditional' sector was undermined by its *contact* with 'modern' or capitalist impulses. For example, the introduction of manufactured goods undermined peasant production; colonial authorities in some areas forced peasants to produce cash crops, which made them dependent on the world capitalist demand and reduced them to 'a branch of agriculture'; colonial authorities forced peasants into the cash economy by taxation and hence created migrant labour, which undermined the pattern of peasant production, and so on. 'Backward' areas are therefore 'bound by fine threads to the externally oriented sector of the national economy, and through it, to the international market' (Slater 1974, p. 333). Indeed, according to some versions, the 'modern' sector grows as a result of its ability to extract surplus from the 'traditional' sector, as well as through trade. Hence the categories of 'modern' and 'traditional' are born in a *single* course of history, and it is the modernity of the 'modern' sector that underpins the backwardness of the 'traditional' sector. For Frank, this process could be metaphorically characterized as a 'metropole–satellite' relationship, in which the metropole draws away the resources and the surplus produced by the satellite. This relationship is a continuous one, with each satellite acting as metropole to another satellite, forming a *chain* of exploitation. In short, dependency theorists argued that, far from a condition of dual economies, developing nations and regions are intricately bound to the global capitalist system, to their perpetual disadvantage.

The assumption of a positive downwards and outwards process of diffusion Significantly, Frank's concept of metropole–satellite relationships is similar in important respects to Friedmann's earlier core–periphery theory,[4] but perceptions of the *consequences* of these relationships are diametrically opposed. In Friedmann's theory, the core is the generator of innovations and developmentally positive impulses which are diffused into the periphery. In Frank's theory, diffusion occurs, but in the reverse direction: the metropole drains the strongest and most competitive factors of production from the satellite, thereby increasing its underdevelopment.

Frank's theory of the metropole–satellite relationship has formed the basis for the reconceptualization of core–periphery relationships by a group of 'dependency' geographers and planners (Soja 1976, Stöhr & Tödtling 1978, Friedmann & Weaver 1979, Weaver 1981). For these theorists, contact between core and periphery is, by definition, negative. This conception has had profound implications for settlement policy. While Frank's metropole–satellite theory is essentially abstract, and refers to economic space, it has been frequently appropriated by dependency geographers and planners to refer to cities, simply because large cities usually represent large concentrations of capital and in this sense can be described as economic cores. In contrast to sector theory, therefore, dependency advocates frequently conceive of large concentrations in cities, and growth poles, as negative. For instance, one view sees growth poles as 'capitalist tools for extending the control of monopoly capital to peripheral areas. As centres of attraction in rural areas they merely intensify the drain of labour and resources from their hinterland' (Fair 1982, p. 21). This idea has been reinforced and expanded by several writers. Lipton (1977), for example, has pointed to the *urban bias* inherent in most resource allocations as a major factor underlying the failure of development plans to bring about more generalized improvement.

The critique of the central pillars of the modernization paradigm was received in varying ways by theorists and development planners. Dependency theorists generally argued for the replacement of capitalist by socialist systems, and in some cases for a withdrawal of nations from an interdependent world capitalist system. Among planners who accepted the critique of the modernization paradigm, the demand has variously been for development based on the specific needs and conditions of developing countries, on internally orientated solutions, on locally defined needs, and on dealing with poverty and inequality in a direct way.

Ecological considerations

A further significant influence on development thinking was a growing concern about ecological conditions. This concern focused upon the use and abuse of resources and the declining quality of the environment, both natural and man-made. These concerns, however, did not manifest themselves in the same way everywhere. One significant school of thought was that represented by the writings of the Club of Rome, which held the position that mankind was on a path to self-destruction. Their first publication, *The limits to growth* (Meadows *et al.* 1972), painted a picture of systematic global resource collapse, but was widely attacked on both methodological and interpretative grounds. Subsequent reports, particularly *Mankind at the turning point* (Mesarovic & Pestel 1974), received less critical attention. *Mankind at the turning point* relocated the problem of collapse to a selective supra-national level. All, however, saw, as the only solution, the development of an internationally managed and balanced, integrated, world economy, operating essentially on the exercise of comparative advantage: a massive vote of confidence in the operation of the multinationals (Friedmann & Weaver 1979).

Another viewpoint, however, was heavily influenced by 'alternative technology' thinking, represented, *inter alia*, in the works of theorists such as Illich (1971), Schumacher (1973) and Turner (1976). Essentially, this viewpoint related the over-development and abuse of some resources to the underdevelopment of other, including human, resources. At the heart of the problem, it was argued, were continually centralizing political and economic systems which both caused, and were increasingly entrenched by, centralizing technologies. These, in turn, selectively and wastefully overused certain resources and underused others. Ultimately, they argued, this approach to resource use would lead to collapse as the centralized systems and technologies dependent upon standardized and non-renewable resources were, in effect, consuming their own life-blood. One important part of the argument stressed *individual* self-reliance: the necessity for people to take greater control of their lives if a significant improvement in the quality of life was to be achieved.

Another theme running through these arguments was the need to reassess the value of resources. Value should be seen, it was argued, not in terms of comparative advantage, but in terms of the absolute value of resources to local areas. By evaluating resources in terms of comparative advantage, the most valuable resources of underdeveloped areas, which were necessary to mobilize less competitive, but locally significant resources, were siphoned off to the strongest areas, thereby ensuring that these local resources remained underused.

The influence of this perspective, therefore, was felt in the demand for 'alternative development' and 'eco-development'; that is, for development strategies with a strong ecological basis.

Major consequences of the forces and influences

In combination, these forces, influences and conditions affected thinking and policy making in the field of development in a number of significant ways.

The meaning and measurement of development

The 'Social Indicators' movement (Seers 1972, Baster 1973) and the redistribution through growth approach of the early 1970s represented the first break with previous growth-centred definitions of development. Seers argued that previously accepted growth-orientated indicators of development, such as gross national product and per capita income, did not give an accurate indication of the quality of life experienced by people, and that factors such as poverty, inequality, employment and housing conditions should be taken as the real indicators of development. Similarly, the redistribution through growth approach to development, while not seriously challenging the priority given to economic growth, argued that the objectives of development should explicitly include reductions in poverty, unemployment and inequality. This trend towards a more humanly based definition of development continued in the Second Development Decade of the 1970s: concerns such as the satisfaction of basic needs, 'liberation from dependency' (Goulet 1979), quality of life, national and individual self-reliance, decentralized and democratic decision making, appropriate technology and the mobilization of local resources became increasingly common in developmental rhetoric.

More recently, the economic crisis in many developing countries has led some institutions to push for an emphasis on 'accelerated growth'. To some extent, aspects of more humanist definitions of development, particularly self-reliance, the mobilization of local resources and 'decentralized' decision making, remain in this approach, but they either exist primarily at the level of rhetoric, or they have very different implications from their original use.

Style of management

One major shift in emphasis in this period has been the rejection by most approaches, to a greater or lesser degree, of the dominant management style of the modernization paradigm. This style, commonly referred to as 'top down', involves the following central ideas:

(a) 'Development in its economic, social, cultural and political dimensions can be generated only by some, very few select agents . . .' (Stöhr 1980, p. 4).
(b) The rest of the population are 'incapable of initiatives in making improvements, consequently everything must be done for them' (Uphoff & Esmain 1974, pp. 28–9).
(c) 'The specific type of development (economic, social, cultural and political) initiated and carried out by the few select agents is the most suitable one for all the others . . . and should therefore replace other existing notions of development' (Stöhr 1980, p. 4).
(d) The type of development initiated by the select agents is not only better than other types, but is also one which others are willing and able to adopt.

The move away from these 'top down' perceptions and towards 'bottom up' approaches also began initially with the redistribution through growth approach. While this approach accepted in a broad sense the idea of external initiators of development, it argued for a development process that would be initiated and carried out by the efforts of the broad mass of people.

While there are numerous conceptions of 'bottom up' development, not all of which are new, it is possible to argue that, as a style of management, it best characterizes approaches developed in the mid-1970s. 'Bottom up' ideas of this period generally reflected the perception that policy interventions in developing countries and regions often destroyed more of value than they provided, not only materially, but also culturally and institutionally (Goulet 1979). Some proponents of the 'bottom up' approach therefore argued that development programmes should be based on existing local institutions, or new ones should be created if they are absent; they should capitalize on 'local knowledge' and should develop existing traditions. Some conceptions took this idea to the extreme of advocating relative autarky and 'territorial development'; that is, development that begins from the resources and needs of a spatially defined region and set of people (Friedmann & Weaver 1979). It was argued that functional integration between regions and nations had led, and would lead, to a weakening of local social institutions, particularly small-scale, informal 'needs satisfaction' networks; to a reduction in the diversity of value systems and to a process of consistent economic underdevelopment of poorly endowed areas which cannot compete in a 'functionally' integrated world (Stöhr 1980). Other conceptions of 'bottom up' lay more stress on the ideas of popular participation and community development, in the belief that a broad-base development programme requires the development of numerous local institutions. This is both for practical reasons and because development is seen as a process whereby people who have previously been marginalized learn to take control of, and hence change, their lives and their collective situation (Goulet 1979). 'Bottom up' approaches, therefore, differ in two main ways from 'top down' approaches. First, development is seen not only as an economic concept, but also as a process dealing with the total human condition. Secondly, proponents accept numerous possible conceptions of development, and argue that development objectives and paths must be appropriate to the context.

It is important to recognize that the concept of 'bottom up' development has also been used by positions which do not challenge any of the central tenets of the modernization paradigm. There are probably two major reasons for this, the first of which is a practical one. Development planners realized increasingly that many problems had occurred in implementing projects because they did not really understand local conditions, social processes and the extent of local knowledge. The idea of decentralizing decision making, and of capitalizing on the existence of local knowledge, provided a way out of this problem. The second reason is an ideological one. The egalitarian and democratic veneer of 'bottom up' rhetoric has suited the ideological requirements of governments and development agencies, particularly the international aid agencies.

The new emphasis, primarily since 1980, of some development agencies on 'accelerated growth' has not meant that the idea of 'bottom up' development has disappeared entirely. Rather, the term has been reconstituted to refer to developments generated through market forces, in the image of *laissez-faire* capitalism, as opposed to development initated by the state.

Notes

1 The Bretton-Woods agreement, by which the dollar and gold were linked, and all other currencies were tied to this, collapsed because the strength of the dollar declined in the late 1960s. Investors became worried about currency devaluations, and put pressure on their creditors to pay back loans. This in turn put pressure on the banks.

2 These include, *inter alia*, the International Labour Organization (the ILO), the World Bank, the International Bank for Reconstruction

and Development, US Aid and the International Monetary Fund (IMF).

3 Significantly, dependency theory itself has been roundly criticized by Marxists, on the grounds that, *inter alia*, it largely ignores class struggle.

4 It should be noted that Friedmann's position has changed radically in recent years.

4 *Changing focus of development planning and implications for settlement planning*

One consequence of the changing conditions and influences in the Second and early part of the Third Development Decades has been the proliferation of different schools of thought about how national and regional development should be promoted. Broadly, four main approaches have emerged.

(a) The first new approach was 'redistribution through growth'. While it accepted many of the central tenets of the modernization paradigm, it argued that a different form of growth, more appropriate to developing countries, should be instituted and that this growth path should be concerned explicitly with issues of distribution. This approach, while widely criticized, formed an important foundation for subsequent approaches.

(b) The second approach, which has come to be labelled 'basic needs', emerged in part as a reaction to the problems encountered in the redistribution through growth approach. There are two main streams of basic needs: one sees a concern about poverty as essentially supplementary to the need to promote economic growth; the other, usually termed 'radical basic needs', argues that a basic needs approach should lay the basis for a *qualitatively* different form of development. Some proponents of this latter approach argue that a broad equalization of access to resources and effective power is a precondition for sustained development.

(c) The third approach, partially related to radical basic needs, but theoretically quite different, argues that development must be *spatially* or territorially defined. This approach – which includes 'agropolitan development' and 'partial territorial closure' – defines the problem of development as promoting 'territorial' as opposed to 'functional' integration, and argues that some sort of withdrawal from the global capitalist system is necessary for development to occur.

(d) More recently, the position that the development of developing nations and regions will occur through increased interconnection into a unified world capitalist system has re-emerged. This position, termed 'accelerated growth', represents to some extent a continuation of the ideas of the modernization paradigm, but with some significant changes in the form and direction of the strategies used to promote development.

Redistribution through growth

The 'redistribution through growth' movement was first heralded tentatively by the 1964 Economic Policy Convention of the United Nations, which formally committed member nations to the goal of full employment, both as an end in itself and as a means of tackling poverty. However, its main momentum was initiated in 1969 with the launching of the 'World Employment Programme' of the ILO, an organization which was to become an important agent in promoting shifts in development policy in the subsequent decade.

While acknowledging the need for growth, the agency recognized that 'part of the difficulty is structural, in the sense that many of these problems will not be cured simply by accelerating the rate of growth' (International Labour Office 1972, p. xi). It initiated a series of countrywide studies to evolve employment-orientated strategies of development. The most important and influential of these studies was the 1972 Kenya report, which formulated a strategy termed 'redistribution from growth'. The approach was later modified and generalized in a seminal work entitled *Redistribution with growth* (Chenery *et al*. 1974) which was produced by the Institute of Development Studies at the University of Sussex in conjunction with the World Bank.

The Kenya report identified a series of structural imbalances in the Kenyan economy which had reflected a series of problems. These were capital-intensive growth in the 'modern' sector which resulted in low levels of labour absorption and extremely low levels of productivity and remuneration in non-formal sectors of the economy, particularly peasant agriculture and the 'informal sector'.

The perceived solution to the problems still rested almost exclusively upon top down management but it now demanded a reduction in inequalities in a number of areas such as income, education and land ownership, and at a number of scales. Further, the interrelationship between economic growth and redistribution was stressed: growth is necessary for sustained, large-scale redistribution; and redistribution is a precondition for sustained growth. In particular, it is necessary to stimulate local demand. The central elements of the strategy, therefore, included a process of income redistribution which would lay the basis for a broader, less import-dependent demand structure; an emphasis on rural development, with particular orientation to the peasantry; a reorientation of the form of industrialization towards processing local materials for local demand; an emphasis on the development of the informal and small-scale sectors, particularly because of their perceived employment potential, labour intensity, low import content and use of appropriate technology; and an attempt to reduce dependence and promote national self-reliance as far as possible.

In a number of respects, the analysis and recommendations of the Chenery group were similar to this. However, the approach advanced focused mainly on increasing the productivity and incomes of the poor. This process might result in a changed pattern of economic development based on broader local demand, but this was not the central focus. Certainly, policies to encourage labour-intensity and limited nationalization were advocated, but ultimately the stress was on the targeted poor: self-employed small farmers, rural artisans, landless rural labourers and members of the urban informal sector for whom income growth was limited by 'a lack of access to land, capital and other public facilities [and] often by outright discrimination' (Chenery *et al.* 1974, p. 15). Further, stress was laid upon the *specificity* of circumstances: precise policies could not be stated *a priori*, although broad policy models for Latin America, Africa and South Asia were defined.

In both forms of the approach (Streeten 1981, p. 182):

> The policy involves taking the extra income that would accrue to the better off and redirecting it to the poor. As proposed, the redistribution would take the form of providing *investment* resources to the poor, so that the redistribution would give them a permanent source of income rather than a temporary increase in consumption. If pursued over a long period, redistribution through growth would, though it started by being incremental, end up by affecting the distribution of income and of assets substantially. . . .
>
> To be successful the strategy requires that the policies will not significantly reduce the growth rate of G.N.P. as conventionally measured. This is of obvious importance to the Kenya strategy since redistribution from growth can only occur so long as there is growth . . . the Chenery strategy redefines the target growth rate to give

higher weight to the growth of incomes of the poorer groups. But if redistributive policies are to lead to a higher growth of this redefined target than the conventional non-redistributive trickle-down strategy, then it is clear that the new redistributive strategy must not too adversely affect the conventional G.N.P. growth rate.

Income redistribution

In the ILO Kenya strategy, the process of income redistribution was to occur through policies of wage freezes at upper income levels and progressive taxation. Finance attained in this way would enable investment in those sectors of the economy dominated by the poor: particularly the informal sector and agriculture. Redistribution of this kind would raise the incomes of the poor, at the same time as reducing those of the wealthy. This would reorientate the economy away from a demand structure based on the production of luxury commodities with a high import content towards the broader development of more basic goods with stronger local linkages, less import content and greater labour intensity.

The Chenery approach was less specific and stressed that redistribution should be limited to a level sufficient to provide the necessary resources to back policies directed towards raising the incomes and productivity of the poor. It argued that excessively high levels of redistribution from the wealthy could reduce savings and capital accumulation, thus leading to lower incomes for the poor. Four possible strategies for redistribution could be used, depending on economic and social conditions:

(a) maximizing growth through raising savings and allocating resources more efficiently, and in such a way as to benefit all groups;
(b) redirecting investment to the poor by access to public facilities;
(c) redistributing income to the poor by taxation or by the allocation of consumer goods;
(d) transferring existing assets to the poor, for example, by land reform.

Rural development

Peasant-orientated rural development is central to redistribution through growth approaches. In the ILO Kenya report, the strategy for growth depended on it. Agricultural development was necessary as a source of raw materials for industry, as a means of satisfying growing demand following redistribution, and as a means of generating foreign exchange. The Chenery strategy saw these as possible outcomes of rural development, but the fact that in many countries most of the population – and most of the poor – lived in rural areas was sufficient cause to necessitate this sort of approach. Rural development would, however, have the added advantage of stimulating demand for the products of labour-intensive small manufacturers. In both cases, a move away from 'urban bias' in services, education and development projects was stressed. This was intended both to remove the 'imbalance' between centre and periphery to curb migration to the largest cities.

In the ILO Kenya report, 'urban bias' was believed to be a 'major force behind migration, both from the country to the town, and from one rural area to another' (International Labour Office 1972, p. 300), and this was a critical factor in problems of urban underemployment, rural decline and in certain structural economic problems. The solution to this 'imbalance' occurred at a number of levels: limits on maximum government expenditure in the largest city and minimum quotas for expenditure in the most marginal regions; regional quotas for access to higher education and to the civil service; raising the income levels of self-employed farmers and other rural dwellers; administrative decentralization; and a policy of dispersed industrialization co-ordinated in space by a system of central places. These measures, it was believed, would 'increase the ability of rural people to lead meaningful lives and thus . . . help to create a balanced rural society from which there is no need to migrate to urban centres' (International Labour Office 1972, p. 314).

Both reports recommended that the rural strategy focus primarily on small-scale, subsistence farmers. Chenery *et al.* (1974) spelt out what this would mean for the archetypal cases of Latin America, Africa and Asia. In the case of Latin America, land reform was critical: this would reduce poverty problems arising from the semi-feudal structure, and would raise productivity. In Africa, it was considered most important to settle and colonize new areas, as well as to provide intensive services to existing small farmers, including stabilization of land holdings, improved seeds and other inputs, extension, credit and market access. For Asia, where agricultural plot sizes were already too small, it advocated the accommodation of a smaller number of farmers on reasonably sized plots. The surplus labour would be accommodated in the long term by a labour-intensive manufacturing process, while in the short term, landless labourers would be given large enough house plots to support limited agricultural activities. In all cases, support activities were to be directed to small farmers to aid productivity increases.

The ILO Kenya report while not advocating the abolition of large farms, emphasized the subdivision and redistribution of many existing unproductive lands. The emphasis on small farms derived from a belief that these would be more labour-intensive and would lead to higher total output and income than would large farms. But if the approach went beyond previous 'growth' assumptions that large-scale farms were more productive, it remained firmly within the growth school by arguing that such peasant development should focus primarily on export and commercial crops. The mission proposed that the concept of 'integrated rural development' be developed and entrenched. This concept is taken up in practice by Chenery *et al.* (1974), particularly for archetypal African cases.

The concept of 'integrated rural development' has subsequently been central to most rural-orientated development strategies, and has had quite specific implications for settlement strategy. The basic premise is that a mutually reinforcing set of inputs is necessary to stimulate rural development. These inputs include the provision of marketing channels, physical inputs into agriculture such as seed, fertilizer, pesticide and so on, equipment for farming, agricultural extension services, credit, adequate transport to markets, roads, repair and maintenance facilities, storage facilities, small-scale processing establishments, appropriate services, particularly functional literacy, vocational training and primary health care. The co-ordination of these inputs takes place through a system of central places or market towns.

Reorientation of industrialization

Instead of industrialization led by import-substitution, which drew heavily on foreign exchange reserves, produced little employment and was dominated by foreign capital, the ILO Kenya report emphasized industrialization on the basis of backwards linkages into the agricultural sector. The advantages of this were seen as twofold: processing agricultural goods for export and import-substitution of basic, essential goods, especially those based on local raw materials. Further, labour-intensive industries, and industries which could be economically located in smaller towns and rural areas, were emphasized. These latter were assumed to be agricultural and small-scale industries.

The Chenery strategy proposed fewer far-reaching changes in

the industrial sector. However, it laid stress on labour-intensity in industry and particularly on small manufacturers who, it was assumed, used more labour-intensive processes than larger manufacturers and produced basic goods for consumption by the poor. In conjunction with this approach, the development of more appropriate technologies which could be used by small-scale manufacturers was emphasized.

Stimulation of the 'informal sector'

Both the Chenery strategy and the ILO Kenya report placed considerable emphasis on the urban informal sector. It was believed that the informal sector consisted mainly of relatively productive, small-scale enterprises characterized by ease of entry, reliance on indigenous resources, family ownership of enterprise, small-scale operation, labour-intensive and adapted technology, and unregulated and competitive markets. It is these characteristics, it was believed, which would make the informal sector the 'Cinderella' of the development process (Leys 1975). Both approaches proposed to increase the sphere of operation of the informal sector by, first, allowing it to exist, and, secondly, by promoting its linkages to agriculture and to the formal sector, for example, by encouraging subcontracting and its use in government projects.

The stimulation of national self-reliance

The perceived need for greater national self-reliance clearly derived from structural economic considerations: balance of payments problems, the capital intensity of foreign investment and so on. However, it also had an ideological element and related to growing demands in developing countries for a New International Economic order. The ILO Kenya report recommended that attempts should be made to curtail the demand for foreign commodities, particularly refined foods, and it held that the state should control the operation of multinational corporations. Here it argued that profits should be taxed to a greater degree, that profit repatriation should be curtailed as far as possible, and that foreign investment should be selected on the basis of whether it would add to the goals of society.

The Chenery group took a slightly different approach. It recognized that inequalities between countries are a major cause of inequality within countries and a major constraint on the adoption of redistributionary programmes. It suggested that international measures should act to support the main thrusts of the overall strategy, and that countries should undertake joint ventures to develop capital goods production in developing countries, in this way reducing dependence. Ideally, it argued, world production should be rationalized in such a way as to enable poor countries to accelerate their industrial development and to enable them to obtain a larger share of the benefits of industrialization.

Implications for settlement policy

Unlike the ILO Kenya report, Chenery *et al.* (1974) spelt out some implications of the approach for the settlement system. In general, they argued for a more decentralized pattern of urban growth, with a system of dispersed small towns serving to modernize and promote the development of rural areas. An approach of this sort had been developed in the seminal work of Johnson (1970), whose central ideas were in line with redistribution through growth approaches. Johnson's work, and the preferred spatial pattern of growth suggested by the Chenery group, formed the basis for much of the subsequent theoretical discussion of settlement policy, with authors such as Friedmann (1974, 1982), Rondinelli and Ruddle (1976, 1978), Friedmann and Weaver (1979) and Stöhr (1981) elaborating on the concept. In many senses, the theory used is not new. It emphasizes the lower end of the urban hierarchy, and most authors use some version of central-place theory to rationalize the spatial distribution of settlements. In practice, however, few of these ideas have been applied in developing countries.

Settlement policy as a mechanism for promoting industrial development Generally, the redistribution through growth approach is somewhat ambiguous about this. On one hand, it explicitly advocates switches in the pattern of industrialization and urbanization towards a more decentralized pattern of manufacturing and settlement. On the other hand, it remains committed to the concept of an increasingly unified world economy based upon comparative advantage, international specialization and division of labour and the central rôle of multinational capital in the development process. This implies concentrated urban development. Although certain authors do advocate interference in the operation of multinational corporations, the measures advanced usually represent minor distributional adjustments to the form of and rôle played by multinationals. While a changed settlement pattern is advocated, therefore, the structural conditions underpinning the highly primate settlement distribution of many developing countries are not seriously challenged.

Settlement policy as a mechanism for promoting rural development The concept of 'integrated rural development', which is advocated in the redistribution through growth approach, has important implications for smaller settlements in rural areas. Specifically, these are seen as *loci* for vital marketing functions and for other essential back-up services, and as sites for agro-industries and other, rurally located, small-scale industries. For rural spatial planners using this approach, these small urban centres are perceived to 'focus' rural development activities, and to provide the central mechanism for the modernization and commercialization of peasant-based rural areas (Johnson 1970, Chenery *et al*. 1974).

The most central function of small towns for the rural modernization approach is marketing. For Johnson (1970), marketing functions in small towns would free peasant farmers from the monopolistic practices of traders which serve to depress prices for agricultural products and hence to discourage production. A central-place hierarchy would ensure that goods produced could be sold at higher levels of the urban hierarchy, thereby extending their marketability and raising prices for the farmer. This would increase the incentive to produce and to adopt more 'modern' agricultural practices.

Small towns contained in a nested hierarchy of urban centres would serve to encourage modernization in other ways as well. They would become supply centres for inputs into agriculture, for credit, for agricultural extension services and for the 'diffusion of innovations'. Small towns would improve employment prospects in rural areas by attracting small-scale industry, thus providing off-season employment and income to agricultural workers. The result, in the long term, would be rural–urban integration and the extension of the national market. The effect of raised rural incomes combined with the development of a hierarchy of urban centres would be to increase demand for products produced in the country, thereby encouraging national economic growth.

Planners working from these premises about the rôle of settlement in rural development have tended to accept the central postulates, and usually the Christallerian form, of central-place theory. Generally they argue for the development of an integrated hierarchy of central places, and give details of functions to be provided at each level; for example, see Friedmann (1974) and Rondinelli and Ruddle (1976). The primary shift in emphasis in settlement planning between earlier growth theories and the redistribution through growth approaches, therefore, is from an 'industry first' to an 'agriculture first' approach.

Settlement policy as a reaction to city size While the negative effects of the increasing size and primacy of large centres in developing countries are noted by proponents of the approach, the speed and scale of the migration process, which is perceived to cause social, economic and political problems in both urban and rural areas, are viewed as the major problems. Johnson (1970, p. 157), for example, comments that:

> . . . more than one Asian or African country is discovering that a planless drift of workers in the prime years of their potential productiveness to sprawling, slum-cursed cities, where huge man-power reserves already exist, may mean not only a tragic misuse of human resources but an equally prodigal wastage of scarce capital by reason of unwarranted pressures on all variety of municipal facilities.

Measures, such as the modernization of agriculture and the spread of smaller agriculturally orientated industries into the rural areas, aimed at improving rural life have, as one important objective, a reduction in migration rates and a change in patterns of urbanization.

At the opposite end of the urban continuum, some authors have, however, doubted the ability of the traditional rural village to act as an adequate vehicle for the modernization process. On one hand, the migration process has drained the villages of their younger and more productive members. On the other, they are physically too small and cannot 'meet the minimum-scale requirements of many modern agricultural processing operations' (Lewis 1962, p. 174). Johnson holds that modernization in these situations calls for 'economic units larger in size than villages' and 'technical, organizational and business leadership which the village landlords, petty traders and usurers cannot possibly supply' (Johnson 1970, p. 169).

Settlement policy as a mechanism affecting service provision Redistribution through growth approaches advocate an improved level of social and utility services in rural areas. It is, however, characteristic of these approaches that service provision is not viewed as a priority of rural programmes, and its functions are rather to correct 'urban bias' in provision and to help raise the productivity of the rural poor. In terms of service provision, education and, to a lesser extent, health are seen as critical to the initiation of development in both the Chenery approach and the ILO Kenya report. However, the Kenya report, in contrast to the Chenery approach, took the position that services other than education and health would emerge primarily as a *consequence* of rural development, a position held in common with the previously discussed growth theories. Further, while the spatial component of these approaches is not well developed, the Kenya report suggests the possibility of service provision through the medium of smaller towns.

The Whitsun Foundation (1980) plan for rural service centres in Zimbabwe illustrates a number of these points. The Foundation viewed the rôle of service centres primarily as one of stimulating economic and social change through the diffusion of 'ideas and innovation'. The function of these centres was not purely economic: the report lists as priority aims, first, rural development, secondly, education, thirdly, housing, and fourthly, health and other services. It is assumed that other services will be attracted to a centre as it grows. Finally, the idea that services should be provided through smaller towns has been used in conjunction with Christallerian central-place hierarchy. It was, however, decided that upgrading selected existing business centres was preferable to the development of completely new centres. The number of rural service centres required was decided on the basis of accessibility – that is, a radius of 10 km was regarded as an acceptable distance for people on foot or bicycle. This gives a service area of some 300 km for each rural service centre and a population of 7 500 or more would be served in areas where densities were 25 persons per square kilometre or more. It is calculated that, on this basis, between 315 and 320 rural service centres would be required to serve those areas in which the minimum density of 25 persons per square kilometre exists. However, it points out that if the shape of the service areas are assumed to be *hexagonal* rather than circular then only 275 centres would be required.

Major criticisms of the approach

An important set of criticisms of the approach are those that led to the development of the subsequent basic needs approaches. In particular, basic needs theorists argued that the approach did not go far enough in providing measures to eradicate poverty.

First, it was argued, too much emphasis was placed on

economic factors. In the process, it 'lost sight of the ultimate purpose of the policies, which is not only to eradicate physical poverty, but also to provide all human beings with the opportunities to develop their full potential'. (Streeten & Burki 1978, p. 412).

Secondly, the policies proposed tended to accept the basic processes of the modernization paradigm. In particular, the rôle of multinational corporations was hardly questioned and the focus in agriculture was on export and commercial crop production. The emphasis on export or commercial agriculture, it was argued, would lead to the relative neglect of food cropping, which was more critical to the survival of the rural poor. Further, an emphasis on export or commercial agriculture would tend to favour wealthier farmers who could more easily respond to incentives to raise productivity in these areas (Lele 1975).

Thirdly, social services provision to the poor was insufficiently emphasized and there was no guarantee that the other policies advanced would ultimately lead to better services being delivered (Streeten & Burki 1978).

Fourthly, the compatibility of policies of 'growth' and 'redistribution' was questioned; it was argued that if there were high levels of redistribution, and investment was redirected to small-scale activities and goods for the poor, growth would suffer, since these sectors lag technologically behind more 'modern' activities. This would limit the ability to redistribute further (Streeten 1981).

Fifthly, a number of assumptions about the effects of strategies were questioned. Streeten and Burki (1978) note that measures such as raising the prices of agricultural products or introducing more labour-intensive techniques have usually tended to reinforce the initial income and power distribution. For example, higher prices for agricultural products led to higher wages which resulted in high prices for products bought by poor farmers. Similarly, Moseley (1978) notes that the assumption that redistribution would result in greater demand for goods produced by labour-intensive and informal sector firms is not borne out empirically. On the contrary, surveys in Kenya and Southern Rhodesia (now Zimbabwe) showed that, overall, increased income resulted in a decline in income spent on informal sector products.

Assumptions contained in the redistribution through growth approach which relate to the informal sector have also been widely criticized. In particular, the approach has been accused of perpetuating a dualistic conception of the urban economy which is generally regarded as inaccurate (Moser 1978). Further, however, the approach assumes a benign relationship between the formal and informal sector: in terms of this conception, the informal sector could be encouraged by promoting closer links between it and formal sector activities, particularly by setting up subcontracting or agency relationships. It has subsequently been argued by some (Tokman 1978) that the relationship between the two kinds of economic activities is not benign, but exploitative. Economic surplus generated in the informal sector is transferred to the formal sector, and resources aimed at supporting the informal sector will simply be transferred to the large economic establishments.

Finally, it has been argued that redistribution through growth is politically unrealistic. As Streeten (1981, p. 163) puts it, 'the required restraint on incomes at the upper income end will be . . . resisted by decision makers who form part of that group'. Leys (1975, p. 264), commenting from an entirely different paradigmatic perspective, concurs. He specifically attacks the proposals of the ILO Kenya report on the grounds of Utopianism and naïvety.

> It saw clearly enough that within the existing socio-economic arrangements the problem of unemployment was certainly insoluble. But its thinking was cast within the logic of a social science whose central concepts ultimately embodied bourgeois interests. What it saw, therefore, was not the contradictory reality, but only an 'imbalance'; not a struggle of oppressing and oppressed classes, but only a series of particular 'conflicts of interests' which the 'leadership' would resolve, if only from enlightened self-interest, in favour of the common good. The mission saw that poverty and unemployment were connected with 'income inequality' and this in turn was linked to the rôle of 'foreign capital' (in the sense of foreign companies producing capital-intensively for narrow markets of relatively affluent consumers). But it did not see that these, in turn, were an expression of, and a condition for, the power structure in Kenya and in the international capitalist system as a whole. They wrote of social or political forces antithetical to their own proposals as 'interests' or 'obstacles' which would have to be overridden or overcome, as if there were some further 'interest', independent of these and more powerful, which would respond to its appeal. But the political power of the compradors, and the political impotence of the 'working poor' were also integral parts of the structure of underdevelopment.

In short, it was argued that the approach essentially required 'the people who had fought their way to positions of power and wealth . . . to agree to surrender a significant part of the advantage they had gained for themselves and their families. . . .

The obvious puzzle presented by these proposals is what incentive the mission thought all these groups – the heart and soul of the alliance of domestic and foreign capital – might possibly have for making such sacrifices' (Leys 1975, pp. 261–2). By definition, it was argued, the only proposals that had any chance of being implemented were those that were not threatening to the power of the ruling classes and it was *precisely* those that were the least effective.

It is significant, however, that the approach taken by Chenery *et al.* (1974) took cognizance of some of these issues. The book contains a lengthy consideration of the conditions under which a redistribution through growth policy might be adopted, and it is recognized that for a number of regimes the strategy is 'out of court' (Chenery *et al.* 1974, p. 72). Redistribution through growth policies, they argue, are likely to be adopted where there is a coalition of interests or a ruling class that sees redistribution as politically necessary. This, of course, will only be true in certain cases. But equally, there are some regimes that have gone considerably further than the policies that the redistribution through growth approaches propose.

Basic needs approaches

By the middle of the 1970s, many development theorists and planners and, in particular, many ILO economists 'had become disillusioned with the redistribution with growth proposals which they themselves had helped to spawn' (Friedmann & Weaver 1979, p. 170). The condition of the world's poor did not seem to be improving and, clearly, poverty needed to be attacked more directly and more radically.

The solution advanced was the basic needs approach. This approach has at its base the identification of a minimum set of goods and services, the provision of which is intended to be guaranteed to a targeted group at the bottom end of the income continuum. The approach stresses that the form in which these goods and services are provided is as important as provision itself, because the form of the delivery process has implications for access of target groups to these services. In general, rural areas are emphasized, although theorists have argued increasingly for a simultaneous *urban* emphasis due to the prevalence of the target group in cities and to their *effective* lack of access to basic needs. A central concern underpinning the approach is that many of the poorest groups are not reached by policies which aim to increase the productivity of the poor – particularly in rural development programmes – precisely because they have no assets.

Diverse interpretations attend the concept of basic needs development (Ul Haq 1981, p. 135):

> To some, the concept of providing for the basic needs of the poorest represents a futile attempt to redistribute incomes and provide welfare services for the poor, without stimulating corresponding increases in their productivity to pay for them. To others, it identifies the ultimate objective of economic development which should shape national planning for investment, production and consumption. To some, it conjures up the image of a move towards socialism and whispered references are made to the experiences of China and Cuba. To others, it represents a capitalist conspiracy to deny industrialization and modernization to the developing countries and thereby to keep them dependent upon the First World. To still others, it is a pragmatic response to the urgent problem of absolute poverty in their midst.

Further, to some it is a complete model, to others an addendum to policy. These various conceptions about basic needs to some extent underline its popularity: it has been capable of being all things to all people, and as a development approach, or set of approaches, it is relatively ambiguous.

Broadly, two main approaches to basic needs can be identified: a more conservative approach, characterized, for example, by the World Bank under the leadership of McNamara, and a more radical approach. In the more conservative approach, the provision of basic needs goods and services occurs in *addition* to existing growth strategies, whereas more radical groups view it as a means of providing *structural change* in developing countries. Despite these differences, there are certain commonalities in the approaches. First, it is agreed that the approach is concerned with removing *absolute poverty* and is orientated to meeting basic needs, defined in terms of the consumption and services of the poorest 40 or 20 per cent of the population: that is, those whose living standards fall below certain minimum criteria. Secondly, basic needs approaches usually define a certain minimum basket of goods and services which should be guaranteed to groups. While it is recognized that what constitutes a basic need will be specific to country and place, and will therefore have to be *locally* defined, certain broad needs are specified. These, theorists argue, have been found to be relatively ubiquitous. In practice, therefore, the following minimum categories are specified:

(a) Basic private consumption goods: adequate housing, minimum number of calories and clothing.

(b) Basic public consumption goods: clean water, sanitation, health, education and, in some cases, transportation.

More radical basic needs approaches also include the objective of *productive employment* at reasonable levels of remuneration and most, but not all, approaches include an emphasis on fulfilling non-material basic needs (Green 1978). The most important of these non-material needs is participation in decision making in those areas that affect the individual's immediate circumstances (Jolly 1977), and this is usually accepted as being central to the provision of basic needs in an appropriate way. Finally, radical approaches stress the need for accompanying resource redistribution, particularly in rural areas (International Labour Office 1976, p. 49):

> . . . partly because it is difficult to confine the effects of standard price, subsidy, tax and expenditure policies to a particular group and partly because the effects of many standard economic policies ultimately are neutralised by offsetting forces set in motion by the policies themselves. For example, the benefits of a rural road-building programme cannot be limited to small farmers; they inevitably 'spill over' or 'leak' to large farmers as well. Similarly, the imposition of a minimum wage for landless agricultural labourers will have multifarious implications for, say, the costs of agricultural production, the demand for labour, the level of food prices and of prices in general – all of which will tend to counteract the effects of the legislation on the distribution of income. Thus, because of leakages and countervailing forces the only way to alter substantially the distribution of income is by altering the distribution of wealth.

In terms of more radical basic needs approaches, meeting basic needs is just one aspect of a process of redistributing wealth and income in order to reorientate the structure of production and consumption (Ghai 1980). This, in turn, would ensure that in the long term, the poor would be in a position to satisfy their own basic needs.

Although the idea of meeting basic needs is more socially orientated in emphasis than are the growth approaches, it does not necessarily deny the need for economic growth. In fact, it is generally recognized that the aims of fulfilling basic needs and growth are not necessarily contradictory, although the precise relationship between the two appears to vary with context (Hicks 1980). Some countries, for example, Cuba, have largely sacrificed growth to concentrate on fulfilling the basic needs of the population. In other countries, where there were small initial inequalities, the fulfilment of basic needs has been combined with rapid growth (Stewart 1980). In still other countries, for example,

China, growth was aided by improved levels of basic needs satisfaction through the development of a healthier, more educated and productive population.

Proponents of the radical basic needs approach, like their counterparts in the redistribution through growth school, argue that the problem of growth is mainly one of *form*. It is argued that if growth occurs from the 'bottom up', that is, on the basis of the improvement in the condition and productivity of human and natural resources in local areas, it is possible to fulfil the demands of both growth and basic needs approaches simultaneously. This approach emphasizes locally made goods, the use of labour-intensive techniques and locally appropriate or indigenous technology (Sandbrook 1982). One consequence of this type of process, which is believed to be conducive to growth, is the envisaged emergence of a new international division of labour, with developing countries producing basic goods and multinational corporations producing only more sophisticated products.

> A focus on meeting the basic needs of people should imply a lessening of the dependence of the Third World on the markets, capital and technologies of the industrialized world; a greater potential for trade expansion among developing countries; an improvement in their terms of trade *vis à vis* the industrialized world; a reduced dependence on and rôle for multinationals and sophisticated technologies; a reorientation of development assistance . . . autonomous, self-sustained growth for the Third World . . . realizing the Third World demands for a restructuring of the world economy (Ghai 1980, p. 45).

More conservative approaches, by contrast, tend to focus on satisfying basic needs as an aspect of social welfare: 'particular policy reforms and projects are designed to assist "target groups" among the poor to escape absolute deprivation; but an unreformed modern sector will continue business as usual' (Sandbrook 1982, p. 8). In the long term, they argue, the main way in which absolute poverty will be reduced will be by increasing the productivity of the poor. In the short term, however, it is necessary to assist people to satisfy their basic needs, for four main reasons.

First, improved education and health are preconditions for increasing the productivity of the poor. For example, healthier farmers can work harder, more educated farmers will understand extension delivery better, and so on.

Secondly, as mentioned, many of the poorest groups are not reached by policies which aim to increase their productivity because they do not have assets, such as land, which are necessary to allow them to enter the programmes. Basic needs programmes

– particularly education and health – will develop such people's only asset: 'their own two hands and their willingness to work' (Burki & Ul Haq 1981, p. 15).

Thirdly, even if it were possible to raise the income of the poorest groups to the extent that they can provide for their *own* basic needs, the market is an inadequate means of supplying some goods, particularly public services. Further, access of the poor to public services is limited, in part because of the way in which these services are defined. For example, health care services are everywhere urban-based, and focus on expensive curative medicine, as opposed to simple, preventive medicine, which is most important in the case of the poor. Public, or private, sanitation and water suppliers concentrate on relatively expensive solutions which the poor cannot afford. Most education budgets are orientated towards urban secondary and tertiary education, but not towards teaching skills that would help the poor in their everyday lives, for example, adult literacy or universal primary education programmes. In short, it is argued by some basic needs advocates that providing basic needs also involves redefining the way in which finance is spent on service provision: standards must be lowered, and approaches that are susceptible to self-help methods and labour-intensive technologies must be employed (Burki 1980).

Fourthly, it will take a long time to increase the productivity of the poorest people to the level that they can afford a full basket of basic needs (Burki & Ul Haq 1981). By extension, approaches that employ basic needs strategies in addition to conventional growth strategies reflect the fear that if patterns of economic development are not imitative of the economic history of developed countries, there is no guarantee that poverty and inequality will be temporary, as they appeared to be in the West (Sandbrook 1982). Further, basic needs approaches reflect the fear that if poverty was not reduced in the 'Optimistic Development Decade' of the 1960s, then conditions would worsen in the face of the slow growth and deteriorating terms of trade which occurred in the late 1970s (Streeten & Burki 1978).

In terms of the *mechanics* of the policy, advocates have been unwilling to prescribe precise directives and have argued instead that the development of such a strategy will necessarily be context-specific. It may be more accurate to characterize the strategy in practice as a set of actions aimed at satisfying particular needs of population, rather than as a comprehensive development approach. This characteristic of vagueness has given rise to a certain ambiguity as to a precise definition of the strategy, and this ambiguity has, in turn, been exacerbated by varied and often confused use of the term. In essence, however, basic needs activities contain a number of identifiable characteristics.

First, employment-generating activities are labour-intensive, use local resources and locally appropriate technologies, and are directed towards meeting local needs: for example, the building of local roads using labour-intensive methods and locally derived materials.

Secondly, the approach emphasizes sectoral inter-linkages. It is recognized that poverty problems are interrelated and that, for example, raising levels of education and literacy will make it possible to use methods of solving sanitation problems that are cheaper, but that require more knowledge and understanding (Streeten 1980).

Thirdly, basic needs strategies are specifically orientated towards the poor. This orientation affects action in a number of ways. One, as indicated above, is the provision of social services which are accessible to people not only in terms of physical delivery – more schools, health care centres, etc. – but also in terms of their content, and the way in which people are treated. Another is that the approaches emphasize development in the *most* marginal areas and of the poorest food producers. Hence, the concept of integrated rural development, popularized in the redistribution through growth approach, is compatible with basic needs approaches. However, in the latter case, programmes should focus on production of *food* – not export agriculture – in part because having enough to eat is itself a basic need (Berg 1980). A third implication of this orientation is that it throws considerable importance on *local power structures*, since these often play a fundamental rôle in determining patterns of poverty or wealth. More radical approaches note that the rich tend to hijack projects aimed at the poor, and that, in practice, it is sometimes impossible to make projects workable without an effective redistribution of assets and power (Sandbrook 1982). More conservative basic needs approaches are somewhat ambiguous on this point: their tendency to concentrate mainly on *delivery* issues to some extent averts the question of local, and national, class issues and the way in which these underlie poverty.

Fourthly, basic needs approaches usually involve an emphasis on local participation in planning (Streeten 1980, Riddell 1977). This emphasis derives from a number of perspectives. One is that most basic needs approaches contain an emphasis on community self-help as a means of meeting basic needs and that this demands local control over decision making. Significantly, the reasons advanced for the emphasis on self-help vary. Some argue that it is cheaper, while for others community self-help is related to the concept of 'development as liberation from dependency': that is, a central part of the *process* of development is the discovery by people that they can take control of their lives (Goulet 1979). In this conception, the practice of local control not only leads to an efficient way of satisfying material basic needs but *itself* promotes the satisfaction of important non-material needs. Another, in purely practical terms, is the importance of local knowledge and local resources in the identification of needs and in the design and implementation of programmes (Lele 1975). It is argued that self-initiated actions by the poor are often logical and creative responses to the social and physical conditions that they face and thus local resourcefulness is a rich resource to be tapped. Further, the tendency by planners in the past to ignore local experience has often been the primary determinant of the failure of many development plans, particularly in rural areas.

Implications for settlement policy

Settlement policy as a mechanism for promoting industrial development Partly as a result of the diversity of possible basic needs approaches, no explicit guidelines for settlement strategy have emerged, but a number of tendencies can be identified. Generally speaking, however, there is a bias away from the view that city-based industrial growth is crucial to the development process.

First, in so far as the approach is focused upon people and not regions, it is not explicitly concerned with interregional disparities. In practice, however, when policies are spatially applied in the case of rural areas, the emphasis is upon the *most* marginal or peripheral areas, since this population, by definition, includes a high proportion of the target population. It must be stressed, however, that this is only an emphasis: there have been arguments, for example, for urban basic needs programmes to be directed towards the major cities and most basic needs strategies include at least the provision of social services in the poor rural areas other than those that are the most marginal (Sandbrook 1982).

Secondly, a bias against large cities is to some extent evident amongst basic needs proponents, but it is not universal (Sandbrook 1982). This derives from a form of core–periphery analysis which relates the problem of poor services and conditions in rural areas to an investment bias in favour of major urban centres. It is argued that measures to improve the provision of, and access to, services, the development of small-scale industry and the provision of support to agriculture in peripheral rural areas will 'reduce the rush to the large cities, economise on the heavy cost of certain services and increase the scope for regional and national participation'. (Streeten 1982b, pp. 3–4).

It is mainly in relation to agriculturally based rural development that industry is considered important. One aspect of this is usually an emphasis on the development of small settlements containing agriculturally related and other small-scale informal industries and providing marketing and service support systems for the surrounding agricultural areas. Given that this industry is to transform the 'structure of production and consumption' and to provide a stimulus for local development, provision for the needs of rural industry – access, electricity, premises, materials and so on – is central to the approach.

Settlement policy as a mechanism for promoting rural development Conventionally, settlement planning for rural development, from this general developmental orientation, is based upon the promotion of small towns. In terms of a radical basic needs approach, the rôle of such settlements would be to focus the transformation of rural life, while for conservative basic needs proponents, small towns would simply provide a 'rational' approach to service delivery. The issue of the support of these centres is somewhat problematic. In terms of the internal logic of the approach, the element of cash earnings through agricultural production assumes less importance than in the approaches discussed previously. Thus, exchange will be limited mainly to the retailing of essential supplementary goods, and, to a lesser extent, the output of small-scale industrial production: this, in turn, will limit the extent to which the settlement hierarchy can be sustained.

Many of the consequences of rural basic needs strategies for the settlement system are indirect and result from the specific measures applied to promote rural development. An important aspect of this is land reform. It is generally held that a broader, more distributed ownership pattern will encourage the development of small urban centres, while a strategy that focuses primarily on *urban* development, at whatever scale, will have little impact on rural incomes if land distribution is highly uneven. Similarly, rural development relies at least as much on the nature or appropriateness of the stimuli provided as upon the way in which they are ordered in space. This realization, in turn, leads away from the dominant rôle which has been accorded to 'urban' elements in the promotion of development.

Finally, it must be stressed that while the emphasis in most basic needs programmes is upon rural areas, this is by no means *necessarily* the case: the emphasis on rural areas has in part been a reaction against the previous 'urban bias' and has been based on the belief that investment in rural areas will bring about higher returns in terms of needs satisfaction than equivalent investment in urban areas.

Settlement policy as a reaction to city size Significantly, the bias against large cities contained in the approach is more related to the *primacy* of large cities, particularly in terms of the spatial distribution of services, than to their size *per se*. This, in turn, derives partly from an acceptance of core–periphery conceptualizations which explain poor levels of services for marginalized populations – which are usually equated with rural dwellers – as the result of 'urban bias' in investment. It needs to be recognized, however, that not all basic needs advocates adopt this position. Sandbrook (1982) for instance, argues that significant numbers of the poor are located in cities and they have little access to services within the city. He argues, therefore, for an approach which, *inter alia*, improves services for the poor in big cities, as these are places where sections of the poor have better chances of survival than in the rural areas.

Settlement policy as a mechanism affecting service provision In contrast with approaches previously discussed, the provision of social and utility services in basic needs approaches is the centrepiece of the development strategy: improved service provision is the foundation for development, and hence it precedes raised productivity, and does not simply follow it, as it does in other strategies.

It is usually accepted that the central issues in service provision are viable and efficient provision on one hand, which in turn requires adequate thresholds, and accessibility to users on the other: an issue which hinges on the concept of range. Most commonly, this conception results in the use of small urban settlements, which are assumed to be accessible to the rural poor, for most service provision and frequently the best spatial organization for these urban settlements is taken to be the central-place hierarchy: for example, see Mayer (1979). In principle, however, the basic needs approach does not necessarily require that servicing points all be *urban*, in the sense that the settlements have functions beyond a few simple services, or that a stereotyped approach must be adopted towards the spatial form of the provision of services. The emphasis placed on determining locally appropriate *forms* and *levels* of service provision to some extent denies the use of a stereotyped approach, and frequently the low level of provision advocated precludes a strong urban emphasis.

Major criticisms of the approach

Criticisms of the approach generally fall into two main categories: criticisms about the assumptions underpinning perceived processes of development, and charges of naïvety and Utopianism. Within this broad categorization, specific arguments require explanation.

Charges related to the assumptions The main argument of the first category is that because of the generality of formulations of the approach, they fail to consider sufficiently processes of development, and that the *implicit* assumptions about these processes are suspect on a number of counts. First, the emphasis given in the approach to the development of the most marginal regions had been questioned. The most marginal areas are not always viable in any terms, for example, because rainfall is too low or the topography unfavourable, and it makes little sense in terms of resource use to encourage the development of inherently inefficient or unproductive activities at the expense of weakening more efficient operations elsewhere. Secondly, it is argued that it is by no means certain that a broader demand structure based on low incomes will necessarily produce a labour-intensive, relatively informal industrial structure using a high local content of materials. For example, poor consumers and producers usually use some imported, capital-intensive products, such as plastic, bicycles, fertilizers and so on (Sandbrook 1982). Further, some argue that tastes have often been moulded to the extent that increased incomes for the poor would probably expand demand for formal sector products far more than for cheap, handmade goods.

Charges relating to naïvety and Utopianism Central arguments in the second category of criticism, relating to naïvety and Utopianism, have largely developed *within* the basic needs approach itself. Radical and Marxist proponents of basic needs have tended to argue that narrowly defined conservative approaches, which are concerned purely with providing basic needs as additions to existing growth strategies, deal with symptoms and not *causes* of poverty. Hence, they are unlikely to lead to the alleviation of long-term poverty (Sandbrook 1982). In particular, it is argued that these approaches are limited as they do not represent a systematic attack on the structures of under-development within a particular context. Radicals argue, therefore, that only a radical approach to basic needs will remove poverty in the long term, and this involves structural economic and political change. Further, it is argued that conservative basic needs approaches are naïve by definition: Basic needs investment in very marginal areas would have to occur over long periods of time before it showed substantial results and this pattern of expenditure flies in the face of the political and economic realities of most countries, where power resides in the wealthiest regions.

The main criticism that is levelled against the radical view is one of Utopianism: it is argued that radical basic needs approaches require certain fundamental political and economic changes to occur before, or by way of, implementation. Specifically, there must be a shift towards an egalitarian income and asset distribution, a move away from technological dependence, and a move towards developing the power of the poorest. In general, therefore, it involves action against existing ruling classes. Why, it is questioned, would it be implemented in a non-socialist society if it constitutes such a fundamental challenge to the ruling classes? One answer advanced is that it may be used by leaders in the form of 'enlightened self-interest' (Nattrass 1982), that is, to legitimate a regime. Arguably, however, measures of this sort are unlikely to go far enough, and would serve mainly rhetorical purposes. This tendency has certainly been notable with regard to rural development policies, which have tended to stay within the realm of rhetoric, or have been implemented in a distorted way, for example, see Dewar *et al.* (1982). In order for a basic needs strategy to have any success beyond simply providing services, the following conditions are probably necessary:

(a) A relatively equal distribution of land.
(b) A democratic local social structure.

Without (a) and (b), rural development policies are likely to exacerbate the position of the poor.

(c) A relatively equal income distribution to change the demand structure. Additionally, however, a reasonable *size* of market is necessary. Considerations of economies of scale have tended to be ignored and examination of the real ability to disperse industry has not been considered adequately. It would, however, appear that the ability of small-scale industry based on appropriate technology to absorb unemployment and to absorb most demand is exaggerated. In China, which is usually advanced as evidence of the viability of the policy, only 10 per cent of industrial production occurred in small-scale dispersed industry (Perkins 1978). Further, the general value placed on small-scale, as opposed to large-scale, production, appears to be largely ideological. In theory, 'small-scale' is valued either in terms of an assumed greater democratic control and/or in terms of the assumed family operation by the very poor. The former is purely an assertion, while the latter will only occur in certain cases, and probably in those operations that are too small to provide a 'motor' to development. Above this very small scale, however, the goal of fulfilling *basic needs* is possibly best served not by specific encouragement to small-scale industry, but by, for example, minimum wage laws or by encouraging effective worker organization. In this regard, international studies have shown that income distribution is positively related to the strength of the labour movement (Sandbrook 1982).

(d) The existence of a peasantry. This condition also, and more strongly, applies to redistribution through growth approaches which emphasize 'back to the land' to a far greater degree. Nevertheless, the importance of the existence of a peasantry is underlined in the basic needs approach by the importance attached to agricultural production in its own right and to its capacity to generate supportive or related industrialization. However, often the condition cannot be met. For example, where capitalism has largely displaced peasant petty commodity production, few 'indigenous' skills are likely to remain. Further, it is unlikely that local industry can or will re-emerge since new patterns of consumption will have been set up. Where penetration has been coupled with extensive migration, as, for example, in large parts of Zimbabwe and South Africa, attempts to recreate a peasantry are likely to meet with failure. In Zimbabwe, attempts to resettle urban migrants, people from overcrowded rural areas and returning bush war fighters on smallholdings have resulted in significant drops in productivity per land unit. While many developing countries have no option but to attempt to settle more people on the land, alternative forms of doing so, which allow far more centralized management at least in the short term, may offer greater prospects of success. Peasant re-creation policies are particularly vulnerable where considerable inequalities in land holdings exist nationally: it may be cheaper for a small farmer to buy food produced by more mechanized processes than to produce it himself.

Agropolitan development and selective territorial closure

Unlike the previous approaches, which were all national strategies with, to a greater or lesser degree, regional implications, agropolitan development and selective territorial closure are essentially regional approaches (Friedmann & Weaver 1979). Further, they are not simply 'regionalizations' of national strategies: they are perhaps best described as attempts to fuse aspects of more radical versions of basic needs approaches with 'territorial' or spatially defined versions of bottom up development. To a large degree, they represent a reaction to, and indeed a direct reversal of, previous spatial policies. Their formulation has been influenced by the rise of regional nationalism in parts of Europe and by the growing impact of the ecological and alternative technology lobbies.

In essence, the concepts have as their intellectual basis an acceptance of a core–periphery conception of underdevelopment. In terms of this conception, the periphery is basically underdeveloped because of its contact with the core and with core region institutions, particularly multinational corporations. Underdevelopment occurs through a number of processes. First, it occurs because of an inevitable transfer of surpluses from periphery to core via banking institutions, taxes, landowners living in 'core' areas, branch plants, purchases of core region products and so on. Secondly, underdevelopment occurs by tying the regional economy to the vagaries of external demand, thereby making the region inherently unstable and subjecting it to unequal exchange. Thirdly, it results from the displacement of local products and activities by manufacturers based in core regions. The solution to these problems is seen to lie in two interrelated actions: the promotion of endogenous regional development on the basis of regionally defined objectives; and protection from the influence of external institutions and demands through selective territorial closure.

The concept of endogenous territorially defined development includes several elements that are seen as both necessary for success and important in their own right: individual and regional economic self-reliance; the development of relative political autonomy; the mobilization of a full range of a region's natural and human resources; the maintenance of regional values and the creation of a regional identity; small-scale economic activity directed towards the satisfaction of basic needs; and the use of 'appropriate' or 'intermediate' technology.

The concept of partial or selective territorial closure derives its intellectual underpinnings from 19th century trade theory: it refers primarily to the creation of selective tariff barriers in order to protect local production. Selective territorial closure is advocated in conjunction with the promotion of endogenous territorially based development in order to achieve a functional disengagement of regions – 'peripheries' – from an integrated international or national capitalist system.

Policies of selective territorial closure are therefore employed to retain regionally produced surpluses locally and this, in turn, is believed to encourage regional diversification of production to occur. Further, closure is intended to protect endogenous industry and to encourage the emergence of a process of import substitution which, preferably, should be based on meeting regional basic needs. It is believed that the process of regional development is aided by the following factors: by encouraging the 'regionalization' of tastes, products and education; by the diversification of agricultural production; and by the development of small-scale industry to process agricultural goods, produce tools for agriculture and to produce goods that will satisfy basic needs (Friedmann & Weaver 1979). The *form* of development is based on community self-help, although this is rarely defined in a precise way. Production for export should only be pursued in products with significant comparative advantage, since it is feared that a more generalized export orientation will result in revenue earned returning to the global financial circuit as 'underdeveloped economies cannot absorb large infusions of capital' (Weaver 1981, p. 97).

In terms of process, these approaches place more emphasis upon the generation and accumulation of surplus than does the basic needs approach. Agropolitan development

> . . . requires that the greater part of any surplus (created through production specialisation within an area) should be invested regionally for the diversification of the regional economy. . . . This process is then envisaged as occurring at successively higher scales. Through retention of at least part of the regional surplus, integrated economic circuits within less developed regions would be promoted and

> development impulses would be expected to pass successively 'upward' from the local through regional to national levels (Stöhr 1980, p. 4).

Perhaps the broadest and, in intent, the most comprehensive conceptualization of the approach, is that of Friedmann and Weaver (1979), elaborated further by Friedmann (1981/2). Clearly influenced by the experience of China, it rests upon two central concepts: life spaces and political community. 'People's life spaces may be viewed as the theatre of their everyday activity. Defined by spatial patterns of social interaction and shared concerns, they are formed because historical patterns of settlement and migration tend to persist' (Friedmann 1982, p. 6). Every individual, household, group and community has a series of territorial life spaces to which they relate: thus, there exists an interrelated hierarchical network of life spaces. 'One such hierarchy includes a person's home, street, neighbourhood, city, metropolis, region, country, multi-country region and the world. There are other ways of arranging and naming it, but some such hierarchy representing the territorial basis of social life is a universal phenomenon' (Friedmann 1982, p. 6). Each life space in turn has 'at least potentially, a political community' made up of those people whose lives are intertwined with that life space. 'This follows from the fact that all life-sustaining activities are centred in a space where they require, for their normal development, a political ordering of social relations' (Friedmann 1982, p. 8). In terms of the approach, territorially relevant and ecologically balanced development is dependent upon decision making in each life space devolving upon the appropriate political community. In a political sense, the nation or region is thus constructed from the bottom up.

At a regional level, the basic unit of organization proposed is an 'agropolitan district' which consists of approximately 20 000 to 100 000 people. This figure is calculated on the basis of an assumed population density of 200 per square kilometre and a radius defined by a day's return journey by foot or cart. The population size seems to have been informed by considerations of adequate thresholds for essential services on one hand, and the need for a size that is small enough to allow for maximum participation in decision making on the other. Ideally, therefore, the nation would consist of a network of agropolitan districts. Within each district, the characteristics of 'development from below' apply: the retention of regional surplus; regional economic diversification and a regional economy directed primarily towards the satisfaction of basic needs; the use of 'appropriate' technology; the stimulation of regional identity; the promotion of local 'territorial', political and administrative structures; local self-reliance and so on. Only marginal resource transfers would take place from outside. Spatially, each district would be organized around a small settlement which supports agriculture and which would be the locus of essential services.

In terms of the approaches, innovations diffuse both from the bottom up and from the top down. The rationale for this lies in the construction of a learning process in which knowledge is seen as accruing not only at the centre but also on the basis of the practical experience and ideas of small groups and institutions. Knowledge developed on the ground in this way is fed into the system as a whole through the local decentralized organizational structure.

There are some differences between proponents of agropolitan types of approaches about the contexts in which they are applicable. For Stöhr (1980), they are only applicable in areas with the following characteristics: a large population, to provide a market; few resources, products or skills which are demanded externally; low standards of living; considerable distance from developed core regions; few dynamic urban centres able to absorb large rural populations; and a specific regional identity. Further, conditions of an equalization of access to land and other resources, a relatively equal distribution of income and the development or re-introduction of democratic, representative organizational structures are necessary.

Friedmann and Weaver (1979), too, specify some conditions for a policy of agropolitan development: the communalization of productive wealth, although precisely what this includes is not defined; equalization of access to the 'bases for the accumulation of social power' (Friedmann 1980, p. 103) – land, water, tools, financial resources, information skills, knowledge and political power; and a facilitative, protective, redistributive and developmentally orientated state. In short, the strategies require, as in China, the development of some form of communal socialism, although how this is to occur is not defined. Further, the approaches can only operate under conditions that are able to sustain fairly high densities of rural population, as the organizational aspects of the approaches demand these.

Subsequently, however, Friedmann (1982, pp. 2–3) has made far more ambitious claims:

> Originally, I had developed the model with reference to rural areas in Asia, particularly in South-East Asia, with their predominant village culture and high demographic densities. Subsequently, I suggested

> that it might also be applicable to parts of Africa where very different conditions pertain. I now think that as a generalised approach to territorially-based development, it is applicable as well to the more urbanised regions of the America's and Western Europe.

Precisely why this should be so is not made clear.

Implications for settlement policy

Settlement policy as a mechanism for promoting industrial development Agropolitan development and selective territorial closure, even more so than the earlier approaches, are influenced by a core–periphery conception which reverses the effects ascribed to the core–periphery relationship as conceived in sector theory. The problems of peripheral regions are seen to stem largely from their overexposure to core-region institutions, values, attitudes and economic activities, either as a result of regional policies or because of the normal operation of the market mechanism. The problems of the periphery can therefore be reduced to too much 'backwash': that is, direct income leakages out of the region through core-region financial institutions; the transfer of profits; purchases made outside the region because of increased accessibility; human 'leakages' in terms of migration out of the region as a result of overexposure to core-region defined education and core-based social and cultural values; economic 'leakages' resulting from the failure of peripherally based economic activities to withstand competition from core-region products; and social 'leakage' or domination as a result of the supercession of regional institutions to the core areas (Stöhr and Tödtling 1977). The agropolitan approach therefore does not so much present a *challenge* to the concepts of sector-derived settlement theory as much as it emphasizes *parts* of the theory (for example, Myrdal's concept of backwash effects) and it reverses the *value* of processes described by the initial theory: under-development is now the result of *too much* diffusion, rather than too little; too much economic and spatial integration; and so on.

In terms of the logic of most forms of the approach, therefore, the region must be protected from outside elements through a selective territorial closure strategy. Major policy elements of such a strategy include:

(a) Reducing core–periphery interaction to the minimum necessary and filtering out the influences of functional core units. In order to do this, the friction of distance must be reformulated from a negative concept 'to a positive one for the structuring of a spatially disaggregate interaction and decision system' (Stöhr & Tödtling 1977, p. 47) and decision-making processes must be decentralized to the local level. The introduction of any new technology or externally based economic activities must be negotiated and the region should be compensated for negative effects. The region must close itself off in terms of production as far as is possible: it can erect tariffs if necessary or institute deliberate buying of regionally produced goods so as to localize capital accumulation.

(b) Introducing a sense of regionalism, or 'regional nationalism', at a cultural and economic level by promoting regionally specific tastes in products, regionally defined education, regional cultural and value systems, regional standards and forms of organization (Stöhr & Tödtling 1977). At an economic level, such regionalism will provide the means to foster self-sustaining local development by a form of import substitution and by product differentiation to protect regional production units from competition. In a broader developmental sense, regionalism will prevent out-migration and will promote a sense of belonging to the region. These conditions, together with the development of local small-scale institutions, will promote conditions of 'loving' and 'being' (Stöhr & Tödtling 1977) and this will assist the fulfilment of social needs.

The core–periphery conception which underpins the approach also affects attitudes to settlement policy. Considerably less importance is attached to the 'urban' as a productive and integrative economic force than in growth-centred approaches. In many senses, the unit of production is not the 'centre' but the hinterland. Urban areas are intended to *serve* their hinterlands by maximizing accessibility within the region. Urban systems are intended to be inward-orientated and relatively decentralized. Hence, the urban-industrial hierarchy would be sustained 'from below' by the relatively stable human, social, political and environmental needs and potentials of its territorial hinterland, rather than by the fortuitous and uncontrollable trickling down of impulses 'from above' (Stöhr 1980). For example, in Friedmann's (1980) proposal for an urbanization policy in Mozambique, the district capitals, which he terms the 'staging areas' for rural development, will grow and develop economically not as a result of stimulation or diffusion from above, but 'as a result of increased rural production and productivity in surrounding areas, particularly in the family and co-operative farm sector' (Friedmann 1980, p. 112).

Development of large-scale industrial activities and urban centres are to be based mainly on regionally defined inputs and demand. Urban centres would develop on the basis of agglomeration economies but would not grow as fast as previously, since their growth would be tied to that of their hinterland, spatially redistributive processes would occur and *economic* efficiency would not be the fundamental rationale of the system. Friedmann (1980) in his proposals for Mozambique, comments that 'growth poles' and 'growth centres' have no worth in a socialist economy since they suggest spontaneous processes of growth and diffusion which would not occur in a centrally planned economy. He does, however, accept the concept of a 'planned territorial production complex' which, through its 'locational interrelatedness' can bring about changes in the existing territorial structure of production and settlement.

The spatial pattern of urban settlements would be decentralized and would maximize internal as opposed to external accessibility. Similarly, the transport network would seek to create conditions of maximum internal, as opposed to external, linkage and accessibility. The settlement hierarchy is seen as the framework for spatial and territorial development and it is integral to the organizational framework of society. Each level of the urban hierarchy is associated with a certain level of 'bottom up' control, as well as with a certain level of functional organization.

Settlement policy as a mechanism for promoting rural development The approaches place considerable emphasis upon this issue. Friedmann, for example, argues that 'to modernise an economy requires as a first step, a strengthened rural base. All other strategies are bound to fail' (Friedmann 1982, p. 24). Part of this process involves urbanizing the countryside in a social sense (Friedmann 1980). The aim is thus to blur the distinction between urban and rural by providing, through a system of small rural centres, urban benefits and services to the countryside. These benefits are more broadly defined than the services identified in basic needs approaches and include access to knowledge, skills, financial resources, information, social networks, production assets, health, education and political institutions. The settlement system, therefore, is seen as a means of providing a rational spatial organziation for, particularly, marketing, dispersed industries and informal sector activities, decentralized administration and political organization, rural support services and social amenities.

Settlement policy as a reaction to city size Although the approaches in general reflect a pronounced bias against large cities, they do not take on the problem of existing large cities directly. Rather, implementation of the approaches is seen as ameliorating the problem automatically in the long term. 'Large cities would not be able to grow as fast as they have in the past – which would help solve a key problem in spatial development in most developing countries' (Stöhr 1980, p. 11).

Settlement policy as a mechanism affecting service provision The provision of essential services is an integral part of these approaches, and, like basic needs advocates, proponents of agropolitan development and selective territorial closure stress the importance of developing appropriate forms of service delivery and of local determination of priorities and delivery design. The spatial organization of service delivery is again based upon central-place theory. 'Inward-oriented urban systems which maximize internal accessibility provide the best conditions for equal provision of basic needs services to all parts of the population' (Stöhr 1980, p. 11). In Friedmann's (1980) proposals for Mozambique, the district capital is the service centre for its surrounding area, with no outlying centre being situated more than 60 km away from it and with two-thirds of the population preferably residing within half this distance. Below the level of the district capital are 'silent centres', providing each village or group of villages – each containing, at minimum, 1000 to 2000 people – with basic services. The centre is 'silent' because it is used during the day, while at night the user population returns to its respective villages.

Major criticisms of the approach

The agropolitan approach, and particularly the more comprehensive form advanced by Friedmann and Weaver (1979), has been widely criticized on a number of fronts.

The first category of criticism relates to problem identification. There are several aspects to this. One is that the core–periphery conception of underdevelopment is, in its own right, highly problematic. Since this conception underpins most of the new development approaches of the 1970s, it will be elaborated on in the next chapter. Another relates to the artificial separation of 'territorial' and 'functional' forces, which in fact comprise a dialectical unity. 'To oppose territory and function is . . . entirely misleading. . . . To assert an eternal meaning to territoriality or to put it against anti-territoriality in some grand schema is . . . pure mystification' (Soja 1982, p. 16). A third relates to the concept of surplus extraction, which underpins the arguments for selective territorial closure. It is clear that, in terms of the approach, surplus extraction from a region is central to the process of underdevelopment. However, the *amount* of surplus which is produced within a region, but which flows out of it, is never quantified. The developmental effects of its retention, therefore, can hardly be judged. It is, however, unlikely that surplus flows out of poor regions are particularly high: least developed – rural – regions operate primarily at a subsistence level and do not produce large amounts of surplus. In those regions where peasant production is largely preserved, for example, and where cash crops are produced for the market, the relation may hold, but again, *how much* surplus is 'extracted' is not quantified. Finally, the implicit assumption that marginal regions would be better off with few or no transfers of finance ignores the fact that some least developed regions are poor because they have few resources and therefore could not develop on the basis of autarky: that is, in some areas colonialists tended to expropriate and develop those regions with best resources – land, minerals, rainfall, etc. – and left the least productive ones to peasant or reserve use (see, for example, the cases of Tanzania, Zimbabwe, Kenya and Zambia in Dewar *et al*. 1982).

A second category of criticism relates to the abstractness of the approach: it fails to come to terms with economic and political realities or processes in any way and is thus 'naïve', Utopian (Sundaram 1980) or merely genteely romantic. Central to this critique is the fact that the approach ignores the existence of classes or conflicting class interests. This weakness, it is argued, would inevitably ensure the implementation of the approach, the emergence of regional élites at the expense of the poor (Douglass 1981). Alternatively, the approach would fail in the face of class conflict. In fact, the approach generally assumes a *tabula rasa* in peripheral areas: there are no marks left by colonial or capitalist penetration and the indigenous society is problem-free. The society is egalitarian and democratic, a coherent community exists and there is neither differentiation within the peasantry in rural areas, nor, implicitly, problems associated with the sexual division of labour.[1] Even as forms of theorizing about alternative strategies for socialist regional development, this approach fails owing to its abstractness and lack of reference to debate about, and experience of, alternative forms of organization in transitional socialist societies (Sayer 1981). Finally, the generalized form of the approach and its analytic vagueness in identifying contextual factors which might influence its implementation, are identified as major limitations.

Unquestionably, there is considerable validity in these criticisms, in so far as they relate to agropolitan development being advanced as a generalized approach to problems of underdevelopment in peripheral regions. This should not obscure the fact, however, that the central *concerns* which underpin the approach are valid in many contexts.

The first is that in many underdeveloped countries and regions, the majority of the poor is engaged in agriculture and that the prospects of significant growth in industrialization, in the face of weak local markets and intense international competition, are remote, at least in the short to medium term. Policies aimed at improving the living conditions of the majority, therefore, *must* primarily, though not necessarily exclusively, engage conditions in the small-scale agricultural sector.

The second is that increasing functional integration nationally and internationally, accompanied as it is by increasing scale of economic enterprise, increasing monopolization and increasing centralization and capital intensification does lead to the systematic overuse of certain resources and the underuse of others. In this situation, regions that are relatively disadvantaged in terms of their resource-base and location find it increasingly impossible to compete.

Thirdly, and related to the above, the output of stronger regions does spill over into the local markets of weaker regions: the flooding of markets that are easily saturated, both agriculturally and industrially, is a major problem to be faced in attempting to stimulate the upwards mobilization of local economies in underdeveloped regions.

Fourthly, an inevitable consequence of functional integration has been the erosion of identifiable regionalisms and the destruction of locally specific social institutions which previously played an important supportive rôle in regional life. Often, too, this has occurred without the creation of alternative institutions, organizations or structures to replace them.

Finally, to a degree at least, the diffusion of innovation is a necessary part of inducing upward mobility in stagnant, underdeveloped regions. One of the most pervasive obstacles to inducing a sustained dynamic in many underdeveloped regions is a shortage of skills. In Mozambique, for example, which sought after independence to build a planned, restructured society, there were recently eight doctors to serve the entire nation of 13 million people; six architects; four planners and a very limited number of engineers. Similarly, towns of 70 000 people had no access to trucks to remove garbage, largely because of the inability to obtain spare parts and the lack of adequate, trained maintenance service. In the face of problems of this kind, Friedmann's concern with learning at the periphery, as constituting the central dynamic-inducing force in development, is understandable, even if his belief that it will simply occur smacks a little of desperation. In the case of Mozambique, for example, some 97 per cent of the population is illiterate and the transmission of knowledge is clearly a massive problem.

Accelerated growth approaches

Recently, concern about the need for redistribution of the benefits of economic growth and the emphasis on the satisfaction of basic needs has given way in certain circles to a re-emphasis on economic growth as the central issue of development. This emphasis, which arguably has never entirely disappeared, is largely the result of the way in which more conservative policy makers have reacted to the economic crisis in both developed and developing countries.

By the early 1970s, most governments in Europe and North America had, to a large extent, abandoned long-term co-ordinated economic planning. Indicative planning had failed to balance supply and demand, as both unions and the private sector proved uncooperative. Moreover, long-term planning became extremely problematic in the face of uncertain and unpredictable world economic conditions. Initially, many governments resorted to short-term demand management techniques as a means of 'crisis management', but by 1975, most governments had introduced austerity measures. State spending, particularly in the sphere of social services, was cut and unemployment levels were allowed to rise. Governments concentrated on improving national economic efficiency and attempted to aid the restructuring of industry.

More recently, these austerity measures have been extended into full monetarist policies in a number of countries, particularly Britain and the United States of America. These policies aim to improve the efficiency and international competitiveness of the domestic economy by, *inter alia:*

(a) Tightening controls over the money supply and pushing up interest rates to encourage efficiency and to discourage unprofitable investments. Tight money controls and the maintenance of a strong currency also act to reduce upward pressures on internal prices and wages.

(b) Reducing state involvement in the economy by cuts in expenditure on social services; pruning subsidies to the private sector; dispersing state shareholdings in companies; and reducing state ownership in the economy. These *laissez-faire* policies are expected 'to achieve the spontaneous regeneration of a vigorous, competitive economy' (Wells 1981, p. 15).

(c) Increasing the openness of the national economy to international market forces. 'The aim is to make international relative prices the standard of reference in resource allocation and to restructure the domestic economy in line with the changing requirements of the international division of labour' (Wells 1981, p. 15). In practice, however, there has been a tendency towards 'new protectionism' (Kaplinsky 1984, p. 77) in developed countries in the face of declining world economic and trade growth. In contrast with older forms of protectionism, and as a result of the commitment to reducing tariff barriers, this has primarily taken the form of 'non-tariff barriers'.

These trends in policy in many developed countries – and also in some developing countries, such as Chile – have influenced policy positions adopted by lending institutions such as the World Bank and the International Monetary Fund (IMF). In recent years, these bodies have advocated policies in developing countries based on a reduction in the rôle of the state and on a coupling of export orientation and openness to trade (Kaplinsky 1984). This direction has been strongly influenced by the rapid economic growth of some Newly Industrializing Countries (NICs) in the crisis years of the 1970s and by a related critique of previous import-substitution industrialization strategies. The major elements of the critique of import-substitution were that it led to excessive government interference, which discouraged private initiative; to a bias against agriculture; to an underuse of labour *vis à vis* capital; and to dependence on capital goods imports and hence to foreign exchange problems. Industrialization based on import-substitution would only allow short-term growth as a result of the small size of the local market and it would be difficult to switch towards export-orientated industrialization at a later stage (Schmitz 1984). As a result, most inward-orientated economies following industrialization based on import-substitution experienced low growth rates in the 1970s. By contrast, most outward-orientated economies following export promotion strategies had experienced high growth rates in these crisis years (Balassa 1984). Increasingly, aspects of the policies followed by these NICs, particularly Singapore, Hong Kong, Taiwan and South Korea, were seen by some as the basis of a new approach to development.

The export-orientated industrialization strategy, as the new approach was termed, stressed rapid, labour-intensive industrialization aimed at export markets on the basis of 'comparative advantage', particularly of cheap labour. This was to be pursued in conjunction with greater reliance on free market mechanisms, the reduction of protective national tariffs and increased integration into the international economy (Beng 1980). State intervention was not entirely dismissed, however and aspects of import-substitution policies, such as subsidies, were retained (Schmitz 1984).

The export-orientated industrialization strategy is not seen as the solution for all developing countries, although important dimensions of the strategy, such as viewing accelerated growth as central to development, the promotion of free market forces, an emphasis on the reduced rôle of the state and an outward economic orientation, are widely advocated. In Africa, for example, the emphasis is placed on accelerated growth centred on agricultural development. The clearest prototype of this approach is the strategy advocated in the 1981 World Bank report entitled *Accelerated development in sub-Saharan Africa* (ADSSA). The report

> . . . was born in response to a 1979 request of the African Governors of the World Bank for a review of the cause and potential cures for their dim economic prospects which they believed confronted their economies. It appeared in the autumn of 1981 as the first comprehensive ideological and programmatic manifesto setting out the post-McNamara Bank's response to lagging development (in many cases disintegration) in the context of rising global economic disorder and deepening industrial economy recession (Allison & Green 1983, p. 1).

The main elements of the strategy are: a reduction in state involvement and activity within the economy; concomitantly, an increase in the sphere of operation of market forces – the report stresses relatively unfettered *laissez-faire* practices; an emphasis on the expansion of export agriculture; a reduction in importance of industrialization led by import-substitution; an emphasis on increasing the openness of the economy; and the promotion of greater integration into the world capitalist economy.

The first emphasis of the strategy, that of reducing the rôle of the state in the economy, is in line with the neo-liberal *laissez-faire* economic policies advocated by monetarist economic theorists. According to these theorists, state management should give way to the operation of a vigorous small-scale, indigenous private sector, and to large-scale private capital, domestic and foreign. The World Bank report lays emphasis on the former and proposes that local entrepreneurs partially replace the state in a variety of areas: particularly in health care and education, but also in the building of infrastructure, marketing, service activities for agriculture and so on. This will occur relatively automatically, it is argued, if the state removes barriers and regulations preventing local entrepreneurs from operating in these areas. Hence, for example, the indigenous trading system can, to a large extent, replace marketing boards and the 'monopoly suppliers of the public sector' (World Bank 1981, p. 65), while public transport can be replaced by private operators and curative public health by private healthworkers who charge a fee for their services. The rôle of the state is primarily one of providing the climate for *laissez-faire* economic growth. In agriculture, for example, it is proposed, *inter alia*, that the state should set pricing structures to provide an incentive for agricultural development, provide infrastructure such as roads to improve market access by traders and farmers, disseminate information on crop prices to encourage competitiveness and manage large-scale grain imports. Hence, while the rhetoric of popular participation is used, the idea of 'bottom up' development refers to a reliance on the market and on small business, though why or how smaller rather than larger businesses should benefit is not defined.

The second major emphasis is upon export agriculture. With this as a basis, the *ADSSA* report stresses the need to increase the output and income of more progressive smallholders in areas of high agricultural potential. It mainly emphasizes pricing and marketing policies to improve the farmers' incentive to produce and it relies on the private sector to provide service and marketing channels in an efficient way. The emphasis on export agriculture accompanies a marked downgrading of the importance of food production. It is argued that local food crop production will benefit from an export emphasis, since services such as marketing, extension and supply, which will grow as a result of export agriculture, will benefit food producers as well. Similarly, it is unnecessary to subdivide large farms, as these 'can be used to spearhead the introduction of new methods' (World Bank 1981, p. 52). Effectively, therefore, the report emphasizes a process of 'trickle down' which is very similar to the assumption contained in sector theory. The authors are not, however, particularly concerned if food crop production does suffer: exports will provide necessary foreign exchange and will stimulate local economic growth (World Bank 1981).

The third emphasis is on further integration into world markets. In this regard, industrialization is seen to occur mainly on the basis of the export of resource-based manufactures, processed agricultural goods and minerals. More industrialized countries which have completed the first stage of import substitution, can develop their more competitive industries further by exporting to other countries. The emphasis upon locally orientated industries, which formed the basis of sector theory, is strongly downgraded. Significantly, the export-orientated industrialization strategy of the NICs, which rests largely on the importation of raw materials for labour-intensive processing by a cheap labour force, is not deemed appropriate for sub-Saharan Africa for two major reasons: wages in Africa are high relative to those in East Asia, and the shortage of skilled manpower forces up costs, making industries in Africa uncompetitive in relative terms (World Bank 1981).

Implications for settlement policy

Settlement policy as a mechanism for promoting industrial development Settlement strategy, on the basis of an accelerated growth approach, can take a number of forms, depending on the context. In general, however, the approach attempts to build on the strengths of the more productive regions and seeks to maximize the cost-effectiveness of investment. From this perspective, investment in marginal areas is only justified where there are underexploited productive resources that can be developed relatively easily or where the investment will stimulate *national* economic growth (World Bank 1981). Marginal regions with little growth potential and without considerable existing infrastructure should be allowed to decline and it is implied that residents will migrate to more developed areas. Central to this position, therefore, is the premise that economic activity does not *necessarily* have to be developed where people live: people can move to jobs. Within this general stance, however, it is necessary to distinguish between three very different applications of accelerated growth thinking. It is important to note that while the three applications are different, they share the common elements of the approach: a commitment – in theory – to *laissez-faire* forms of growth, a reduction of the direct rôle of the state, an emphasis on deregulation rather than direct subsidy as the primary form of incentive and a commitment to export-orientated growth.

The first application occurs in the context of more developed countries. The clearest example of this type of thinking can be found in the concept of free enterprise zones, which are intended to develop a vigorous competitive sector in parts of cities that have 'been deserted by the collapse of the postwar capitalist boom' (Massey 1982, p. 433) and in cities that have been losing their relative industrial dominance for some time (for example, New York). Spatially, therefore, investment is applied to large but old cities that are relatively stagnant economically.

In theory, free enterprise zones are intended to replicate conditions found in places like Hong Kong and Singapore. They would be outside the range of taxes, payment for social services, industrial health and other regulations, minimum wages and exchange controls. The zones would be intended to attract small firms, which, it is believed, have a high innovative and employment-generating capacity. Hence, they would progressively increase the sophistication of their operations from low-level developments involving few technical skills to becoming independent centres of technological innovation. In the process, older skilled workers whose jobs have been de-skilled and/or lost and young unemployed workers would be employed and retrained in new, more marketable, skills.

In practice, the proposals are not so much a revival of *laissez-faire* capitalism as a further means of subsidizing capital to locate in desired places (Hall 1982). In Britain, benefits for location in enterprise zones include a ten-year real estate tax exemption and a tax allowance for full investment depreciation. Some planning controls have been removed, investors are exempt from land development taxes and are required to file fewer statistical reports (Goldsmith 1982). In the United States of America, the enterprise zone proposal eliminates business capital gains taxes and provides other tax credits. Employers receive tax credits for each low-wage, unskilled or black resident they hire (Harrison 1982). Favoured locations for enterprise zones are those cities that privatize and cut public services, prohibit rent control, minimize regulations and weaken unions (Goldsmith 1982). The free enterprise zone concept has also been used to attract industrial development in a number of Newly Industrializing Countries, such as Mexico, and it is increasingly advocated as a means of stimulating industrial development in poor countries and regions. The use of the concept in the extremely marginal Ciskei Bantustan in South Africa is an example of this (Jones 1984). In a number of ways, the use of the free enterprise zone concept is similar to the use of growth poles, except that different types of incentives are used: incentives relate to reduced controls, such as the relaxation of regulations protecting workers, tax exemptions, abolition of licensing and zoning requirements, and the like, rather than direct subsidies.

The second spatial form which the accelerated growth approach has taken is the channelling of investment into the most productive and therefore the fastest-growing areas. These areas are usually large cities which are well linked internationally, where production costs are low and where cheap labour is abundant. This type of policy is, to a large extent, an effect of export-orientated growth approaches.

The third application of the approach is in peripheral, predominantly agricultural, regions and nations. Here the emphasis is on resource-based industrialization and on the modernization of agriculture. This will be discussed further in the next section. Suffice it to note here that the emphasis in relation to the modernization of agriculture is upon those small and medium-sized towns that have strong communication links to export markets.

While the emphasis of the approach is squarely upon the promotion of economic growth, the significance of the diffusion of the benefits of growth is not totally denied. Rather, the approach implicitly accepts the basic premises of sector theory: that growth is necessary *before* diffusion or redistribution can occur and that diffusion will automatically occur downwards through the settlement hierarchy. However, whereas sector theory saw a rôle for policy in promoting an integrated hierarchical system of settlements to assist innovation-diffusion, the accelerated growth approach *implicitly* holds the position that the settlement system will move towards the 'best' form automatically through market forces: artificial attempts to adjust the hierarchy before the market is ready simply represent wasteful investment. To a large extent, this view duplicates the position of the automatic-adjustment thesis of the 1960s, which argued that high levels of spatial inequality are a feature of developing countries and that 'convergence' and an integrated settlement hierarchy will result once a certain stage of economic development has been reached. In practice, however, elements of settlement policy from previously developed approaches, such as the stimulation of small town development or the promotion of a hierarchy of central places, are sometimes used in conjunction with the approach. As Zingel (1984) points out, the promotion of a central-place hierarchy in the context of a free market approach in the Ciskei Bantustan of South Africa is a contradiction in terms. On one hand, free market policies are unlikely to develop the rural areas – which are currently highly marginal in terms of agricultural production – in a way such that they can support low-order centres. On the other hand, industrial and commercial establishments are unlikely to be attracted to centres designated as medium or low order in terms of the hierarchy, as there is too little income to support them in these locations. Similarly, while the stimulation of declining cities in some developed countries is in contradiction to the automatic-adjustment thesis, it is accepted that large cities which have considerable existing infrastructure cannot simply be discarded.

Settlement policy as a mechanism for promoting rural development A central focus of the accelerated development approach in least developed areas is 'rural modernization' through the stimulation of export agriculture. It is argued that services and marketing functions generated around, and by, export agriculture, will also serve food producers. Although *ADSSA* makes little direct reference to settlements, in general the development of small and medium urban centres is seen as a vehicle for developing export agriculture, though *how* these are to develop is unclear. Presumably, since the approach favours a market-centred orientation, small centres are expected to emerge or develop autonomously. Small centres are believed to play a number of rôles in the promotion of export agriculture.

First, by causing a build-up of thresholds, they provide efficient locations for the collection, marketing and distribution of produce. It is the marketing function of small towns in particular which, it is believed, will stimulate productivity increases. Secondly, it is argued that the inducement to buy the types of goods found in urban centres will encourage peasants to move into cash farming. Thirdly, it is envisaged that small urban centres will in the long run, absorb population from the land. Fourthly, small centres provide a locus for the efficient provision of financial, commercial, social, technological and utility services, though the provision of these functions will not be guaranteed by the state. Fifthly, they form important links in the innovation–diffusion chain, which is still regarded as important in the long term. The *policy* implications of the strategy are, however, unclear. *ADSSA*, for example, makes no reference as to how these centres will develop, beyond laying stress on the development of communications and particularly feeder roads which, presumably, will stimulate small centres by facilitating access of rural producers and consumers to markets located in these centres.

Settlement policy as a reaction to city size Theoretically, since accelerated development approaches stress efficiency, they should be entirely compatible with the promotion of large centres, since profitability and productivity are generally higher in larger rather than smaller urban centres. Moreover, the emphasis on export promotion would tend to encourage the growth of fewer large centres. In practice, however, accelerated growth theorists do not automatically condone the unfettered growth of the largest centres. *ADSSA*, for example, advocates that measures should be taken to curb the size of the largest cities, mainly because of the potential political threat which a concentration of many under-employed or unemployed people represents.

Settlement policy as a mechanism affecting service provision In terms of the logic of the approach, social, commercial, utility and other services are, as far as is possible, provided through the market mechanism. The state retains the rôle of providing or initiating the provision of certain basic welfare services such as communications, basic education, preventive health care and low-cost water and sanitation. While the state would therefore have to choose the location of welfare services, most other services would find their own 'logical location'. Since 'logical location' in this sense is determined by the spatial distribution of the market, the approach is, with respect to its service provision, entirely compatible with central-place theory. In rural situations, where there is a more or less even distribution of population of sufficient density to support a series of settlements, the logical pattern is also consistent with the Christallerian *form* of the theory.

Major criticisms of the approach

The approach has been criticized on a number of grounds. First, it has been argued that the assumptions upon which the approach is based are incorrect. In particular, two arguments which have been advanced as important justifications for the approach have been refuted by critics. One is the claim that governments in developing countries, particularly in Africa, are inefficient (Colclough 1983). The other is that there is limited state involvement in the more successful NICs (Beng 1980). In South Korea, for example, 'the state has played an active and central rôle in the allocation of resources' (Schmitz 1984, p. 12). It has placed restrictions on direct foreign investment, introduced tight discretionary control over investment and export decisions and has made access to the protected domestic market conditional upon satisfactory export performance. Similarly, it is argued that the earlier experience of industrialization led by import-substitution provided an important *basis* for export orientation in the NICs (Schmitz 1984). Further, Balassa's (1984) definition of countries which are 'outward orientated' is problematic and this undermines his correlation of outward orientation with growth and therefore his argument in favour of export-orientation (Fishlow 1984).

Secondly, critics stress the *specificity* of the NIC experience. It is argued that the success in growth terms which was achieved in Singapore, South Korea and Taiwan derives from their specific circumstances: historical, locational and political (Goldsmith 1982). In South Korea, for example, some of these specific elements were: the radical land reforms carried out in the late 1940s and early 1950s which allowed the state to develop agriculture on the basis of a strong smallholder system; a strongly interventionist state which played a strategic rôle in the formulation and implementation of development strategy; an early experience of industrialization based on import substitution which 'played a key rôle in laying the foundation for Korea's development of an industrial working class and the beginnings of an industrial class in spite of Japanese control' (Evans & Alizadeh 1984, p. 33); the achievement of universal primary education and adult literacy by the 1960s; the inflow of entrepreneurs, professionals and technical personnel following the Korean War; and the country's strategic geo-political position, which has affected the extent and the form of foreign aid (Evans & Alizadeh 1984). Similarly, the other NICs have had highly specific developmental experiences. The validity of attempting to follow a generalized development approach which has only been successful in a few cases is thus seriously questioned (Harrison 1982, Massey 1982, Goldsmith 1982, Evans & Alizadeh 1984, Schmitz 1984).

This point about the viability of the approach in different contexts has been levelled at all forms of the approach. In terms of developed countries, it is argued that the use of free enterprise zones as an attempt to develop a vigorous competitive sector in blighted inner city areas is naïve. Zones will favour large established enterprises, as one of the basic incentives necessary for small firms – venture capital – is not part of the package offered in these areas. Even if small, labour-intensive industries do establish themselves, experience elsewhere has shown that they will not necessarily become more diversified or provide more highly skilled jobs at a later stage. Hence, the zones – if they work – will result in an increase in the size of the marginalized, unstable work force. These problems will be exacerbated in the rest of the city as a result of more profitable conditions for business in the enterprise zone (Massey 1982, Goldsmith 1982, Harrison 1982): critics argue that the enterprise zones may simply blight other areas by attracting industry from elsewhere, while not creating new jobs *per se*.

In terms of most developing countries, it is argued that there is limited scope for the labour-intensive export form of industrialization which is occurring in the NICs (Goldsmith 1982). As Cline (1982) points out, if a large number of developing countries were to pursue this strategy at the same time, their output would be more than developed countries could absorb. This would lead to protectionist responses on the part of developed countries which would make the strategy untenable. This issue is particularly important given the weak state of the world economy and that the East Asian economies adopted their strategies when 'the world economy was in a phase of prolonged buoyancy' (Cline 1982, p. 89). In fact, a tendency towards protectionism is already noticeable (Beng 1980). This is affecting in particular labour-intensive sectors, in which developing countries are being encouraged to specialize, as well as those firms that are not subsidiaries or affiliates of firms in developed countries (Kaplinsky 1984).

The viability of the accelerated growth approach in peripheral rural areas has also been questioned. On one hand, the prospects for rapidly growing, export-orientated, resource-based industries are poor, as the international demand for these manufactures is growing slowly (Godfrey 1983). Prospects within regional markets are also limited and are restricted to only a few economically stronger countries. On the other hand, the slow growth of trade restricts the possibility of achieving rapid economic growth through export agriculture, while the risk involved in neglecting the food-producing sector is enormous. As Godfrey (1983, p. 43) notes, the policy could be a 'recipe for mass starvation'. The specific proposals of *ADSSA* have been criticized on other grounds as well: Green (1983), for example, argues that the strategy is unrealistic in expecting governments to raise domestic prices while world purchasing power is declining; it is narrowly economistic in neglecting social services which are important in raising productivity; and it is vague in that there is 'no coherent presentation of a consistent priority package' (Green 1983, p. 31).

Finally, the approach has had levelled at it all the major criticisms that the dependency theorists brought against the growth paradigm, particularly the assumption of a relatively automatic trickle down of economic growth. Even if growth is achieved, it is claimed, there will remain a significant proportion of people whose condition will not necessarily be improved. Indeed, it will probably be worsened as a result of the withdrawal of state-provided services (Allison & Green 1983). In short, it is widely argued that 'this type of model is not only socially unfair but it also contains economic and financial contradictions which increasingly undermine its viability' (Griffiths-Jones 1983, p. 53).

Recent attempts to mesh policy approaches

Much of the previous discussion has attempted to outline the implications – either explicitly articulated or implicit – for settlement policy of different approaches to stimulating national and regional development. More recently, however, several authors, such as Richardson (1981), Renaud (1981) and Gilbert and Gugler (1982), have been calling for more pragmatic approaches to development policy in developing countries. Although there are significant differences between them, this new pragmatism is generally underpinned by a number of broad areas of agreement.

First, they accept the validity of both growth and distributional goals. On one hand, there is a rejection of a crude dependency based position that automatically correlates economic growth with underdevelopment. It is accepted that, in poorer areas, and particularly those with a relatively rapid population growth, there is an unquestionable need to achieve more rapid rates of economic growth. On the other hand, there is agreement that economic growth is not sufficient and that its increasingly uneven nature, internationally, nationally, regionally and interpersonally, is profoundly problematic. The call, therefore, is both for increased economic growth and for greater social justice, in the sense of achieving a more equitable spread of the benefits of growth and of improving levels of welfare in developing nations and regions. This position represents, in part, an attempt to mesh concepts of both the growth or 'top down' approaches to development and the more recent 'bottom up' approaches.

Proponents of this eclectic position do not define 'bottom up' very precisely. However, the term is generally held to refer to those models, approaches and attitudes to development which approximate 'development from below' and which *inter alia* include an emphasis on development in the spatial periphery, autarky at some level, egalitarian development and 'bottom up' styles of management. In policy terms, it represents an attempt to mesh state participation and its priorities with those of particular districts and with peasants' needs (Friedmann 1982). It advocates the use of a variety of policy forms where they are perceived to be most appropriate. Neither the 'top down' nor the 'bottom up' approach, it is argued, is totally correct or incorrect. It is held, for example, that 'top down' strategies such as growth poles are not *inherently* negative. In many cases, the failure of these policies was caused more by inappropriate and unimaginative implementation, a failure to sustain the policies for long enough and a lack of supportive policies (Hansen 1981), than by inherent weaknesses in the approaches. Their successful implementation demands that they are only employed to achieve limited and feasible objectives and that they are implemented only in those places where they have a reasonable chance of success. On the other hand, they argue that many of the 'bottom up' strategies are impractical, given the political structures of most developing countries. Further, they claim that autarkic solutions are likely to take as long to bring about regional economic convergence as 'top down' approaches (Hansen 1981). Significantly, however, 'bottom up' issues are not entirely rejected. Thus, for example, the rôle of small settlements in stimulating 'bottom up' development and, indeed, 'bottom up' approaches to the stimulation of these settlements, is considered to be important.

Secondly, the importance of both aspatial and spatial policies, and the need to integrate these, is emphasized. Generally, a cautionary note is sounded about an over-emphasis on spatial policies. This position is based on two main arguments. One is that social and economic policies contain *implicit* spatial biases, which, in many cases, have tended to operate in contradiction to the objectives of settlement policy. For example, industrialization led by import-substitution favours major urban centres, whereas sectoral biases to agriculture will encourage the development of small urban centres (Renaud 1981). An emphasis on aspatial policies, they argue, may be a more effective way of fulfilling social goals, *as well as* influencing settlement patterns. For example, the goal of raising rural incomes might be served best by encouraging migration to urban areas (Richardson 1981), and that of reducing urban growth might be achieved best by attempting to reduce fertility. Similarly, it is recognized that a highly skewed land distribution pattern will result in most of the benefits of rural development policies being felt in the largest cities and not in designated small towns. The other central argument is that spatial policy is a highly ineffective means of pursuing goals of equity: policy in this regard should be direct and not rely upon 'trickle-down' effects (Renaud 1981). Consequently, there is less emphasis on the concept of the 'urban' as necessarily being developmentally positive and expensive policies – for example, new town developments – in peripheral areas without a specific resource base are seen as wasteful. It is important to recognize that these arguments, which stress the significance of aspatial, primarily economic, policies are not new: indeed, most 'bottom up' strategies have been of this kind and settlement policy has in most cases either been vague, partially supportive, or tacked on. In the implementation of policies, the split between spatial and aspatial policies has often been maintained deliberately because of political considerations. A common example of this is a strong rhetorical commitment to rural development, often accompanied by a certain amount of investment in infrastructure, while a pronounced urban bias in investment is maintained.

Thirdly, in relation to settlement planning, a cautionary note is sounded about isolated actions or the overstressing of any one part of the settlement hierarchy. It stresses the need to view the settlement system as a cohesive whole, which needs, in policy terms, to be considered as a totality in the form of a national settlement strategy.

Fourthly, these more pragmatic approaches tend to lay great stress on the importance of *contextual* factors in policy design. Factors such as the size and shape of a country, its topography and climate, the form and level of economic development, the level of urbanization, political structures and 'cultural heritage' are important (Richardson 1981). Contextual differences, as well as differences in terms of social goals, lead to a variety of different potential types of national urban development strategies. Richardson (1981), for example, defines ten such strategies: *laissez-faire*, polycentric development of the primate city region, leapfrog decentralization within core regions, counter magnets, small service centres and rural development, regional metropolis and subsystem development, growth centres, development axes, provincial capitals and secondary cities. None of these, he argues, constitutes a national urban development strategy on its own, as all have a limited sphere of operation. For example, it is held that attempts to accommodate *all* rural surplus population in small service centres will be more costly than a more concentrated pattern of urbanization, but the use of small service centres which play a limited rôle in rural areas is recognized. Thus service centre strategies could be combined with some other strategy to deal with the distribution of population at higher levels of the urban hierarchy or within a primate city region.

One pragmatic approach which, to a greater extent, has the characteristics of a 'blueprint' and which places settlement planning centrally in the development process, stresses the rôle of secondary cities in promoting development.

> An underlying assumption . . . [of the position] is that a system of functionally efficient intermediate cities linked to larger and smaller urban centres and to a network of rural service and market towns can make an important contribution to achieving widespread economic growth and an equitable distribution of its benefits in both capitalist and socialist societies. [The approach] . . . recognizes the need for national governments to commit their substantial resources to strengthening intermediate and small-scale cities through 'top down' planning as well as the need for strategic investment in social and physical infrastructure in rural market towns and villages to facilitate 'bottom up' development (Rondinelli 1983, pp. 12 & 15).

The approach has been endorsed by a number of influential lending agencies such as the World Bank, and secondary city programmes have been implemented in a number of countries since 1972, including South Korea, Brazil, India, Mexico, Nigeria, the Philippines, Indonesia and, recently, in Kenya and Thailand. Because of this, it warrants more detailed attention.

Central to the approach is a concern about the growth of the primate cities in less developed countries. One aspect of this concern is a perception of growing social, economic, physical and, importantly, political problems within the primate cities themselves. Another aspect is the degree to which primate centres appropriate national resources.

> Primate cities and large metropolitan centres have concentrations of national resources and social overhead capital vastly greater than their share of national population and from which only a small percentage of the nation's people obtain direct benefits (Rondinelli 1983, p. 29).

A third aspect stems from a core–periphery conception, which views the migration of people and resources to the metropolitan centres as a major cause of underdevelopment in the peripheral rural areas. Even where problems in the primate city are not particularly severe, proponents of the approach argue for an adjustment in the settlement hierarchy to be initiated *before* many of these problems emerge.

The emphasis on secondary cities – defined as all urban places which are other than the main urban centre; which display functional complexity; and which, in Rondinelli's (1983) definition, have populations in excess of 100 000 people – stems from a recognition of the significance of economies of agglomeration and scale in determining spatial patterns of development and from a recognition of the need for efficiency in resource use. An important dimension of this concern for efficiency relates to the efficient use of limited technical, professional and administrative skills in the face of the ubiquitous skills shortage confronting developing countries. Significantly, the approach specifically rejects the 'automatic-adjustment' assumption underpinning some conceptions of rank-size theory. It argues for a conscious strengthening of these secondary cities through co-ordinated policy packages.

Finally, it is argued that it is in the secondary cities that the urban and rural poor have most contact and where their relative conditions most strongly interpenetrate. The focus of policies on secondary cities, therefore, is compatible both with a targeting approach directed towards the poorest groups and with the goal of increasing income and welfare convergence. Indeed, it has been suggested that a secondary city development strategy, '. . . by emphasising the importance of *place* rather than *programme*, is a significant and innovative shift from the conventional emphasis on the need for planners to meet basic needs largely by throwing money at them' (Rondinelli 1983, p. 40). Significantly, the primary emphasis is upon the use of secondary cities to stimulate rural, as opposed to industrial, development, although industrialization is not entirely ignored. Specifically, the centres are supposed to fulfil the following rôles: encourage the commercialization of agriculture by providing markets and improved marketing and by providing the necessary back-up services for increased production; promote the provision of better services in rural areas; provide opportunities for employment – sometimes only seasonally – through decentralized job opportunities and the promotion of agro-industries; assist the integration of poorer regions and groups into the national economy; diffuse social and technological innovations to underdeveloped areas; and promote increased regional convergence of incomes and welfare.

Policies to develop secondary cities should be directed towards four main objectives: extending support services for economic development; improving physical infrastructure; strengthening the economic base and employment structure; and building the planning, administrative and financial capacity of local governments (Rondinelli 1983).

Note

1 See, for example, Henn (1983) for details of how this affects productivity in Africa.

PART III

Differences and debates relating to settlement policy

5 *Major debates*

Previous sections have outlined the implications of different approaches to national and regional development for the settlement system and for settlement-directed policies. This chapter attempts to summarize, compare and expand these implications by focusing on some of the major debates relating to settlement policy. The debates fall into three broad categories: debates relating to the largest urban centres, debates relating to an integrated functional hierarchy of settlements and debates relating to small settlements.

Debates relating to the largest urban centres

In recent decades, there has been a growing concern, particularly in developing countries, about the growth of the largest cities. The issues underpinning this concern are of three interrelated but distinct kinds: those related to the *size* of the city, those related to a *core-periphery* conception and those related to the *primacy* or excessive dominance of the largest centres.

City size debates

In most approaches to national and regional development, the issue of size is not raised directly: all, however, are informed by implicit attitudes to size.

Export-base and sector theories, which constitute the most common forms of the modernization or growth paradigm, argue that large cities have a positive initial rôle to play in promoting generative developmental impulses. Implicitly, they suggest that the growth of large cities will be self-correcting in the long term: as the settlement hierarchy develops there is a move towards equilibrium in population distribution and migration will be reduced to a minimum. The approaches hold that, in the second stage of development, in particular, it may be necessary to introduce policies to assist this process and to reduce pressure on the largest cities. Sector theory, in particular, is entirely compatible with the creation of growth centres, new towns and deconcentration points, one function of which is to slow down the growth of the largest cities.

Most of the other approaches discussed reflect, to a greater or lesser degree, a bias against large cities. Although there is no inherent reason, in terms of their internal logic, why accelerated growth approaches should contain such biases, even the World Bank-formulated accelerated growth approach for sub-Saharan Africa (World Bank 1981) advocates measures to control the growth of the largest cities, primarily for political reasons. At the same time, however, proponents of the accelerated growth approach argue for measures to *cope* with rapid urban growth, largely for pragmatic reasons.

The main arguments against the growth of large cities are of two kinds. One, informed by dependency theory, relates to the impact of these large cities on less developed regions and settlements. Essentially, it is argued that this impact is negative, exploitative and results in accelerating urban bias in investment. The other relates to conditions within the cities themselves. It is held that with increasing size, diseconomies such as rising land values, rents, transport costs, crime, pollution, noise and so on, outweigh economies of scale in many cases. Even if these diseconomies do not outweigh them in absolute terms, they often affect the poor disproportionately. When viewed in welfare and not aggregate terms, therefore, the situation is worsened by increasing size (Gilbert 1976).

This reaction against large cities, however, is by no means universal. In practice, many agree with Richardson (1972, 1976), Fuchs (1967), Alonso (1972), Mills (1972), Hoch (1972), Wingo (1972) and Mera (1973) that, particularly in developing countries, the benefits of the economies of scale, that can be gained in large cities outweigh the diseconomies that set in with the attainment of large city size. In policy terms, they question the wisdom of attempting to curb the growth of large cities and of dispersing urban industrial functions in particular to smaller towns and cities in the periphery. Rather, they argue, in situations where capital, resources and skills are scarce, it is more efficient to concentrate investment and to capitalize on the economies of scale thus provided (Alonso 1972). These arguments have fallen on receptive ears in a number of international lending agencies. Gilbert (1976, p. 27) summarizes the position:

> It is ironic, of course, that these studies should appear now when the wisdom of decentralisation has never been more widely accepted. Policies to build new towns and to reduce unemployment and raise

> income levels in poor regions have been common in the United States and Western Europe since the 1930s, but only since the early sixties have they been adopted widely in other parts of the world. In Latin America, years of persuasion by foreign planners have finally convinced governments that they need regional policies: Mexico, Brazil, Venezuela and Colombia are encouraging industrial decentralisation and the growth of intermediate cities. . . . In Africa, regional and rural development policies are no longer unfashionable and a country such as Tanzania has adopted strong decentralisation policies and is contemplating building a new national capital. The new studies in favour of large cities, therefore, threaten a conventional wisdom which has taken many years to emerge. The process of erosion is, however, already under way. The World Bank has contracted studies into the feasibility and desirability of large city development in Latin America and other parts of the Third World. Planning teams have already produced reports predicated on the belief that cities of eight million people or more should be allowed to grow.

Four main arguments are advanced to support this position. First, it is argued that the poor benefit as a result of increased city size. Per capita incomes rise in direct proportion to city size and even where high urban unemployment exists the survival chances of the poor are greater because of increased participation in the informal sector, access to formal or informal support systems and so on. Secondly, large cities generate faster national economic growth rates. Thirdly, the costs of social overhead capital tend to fall with increased city size. Finally, it is argued that while problems of urban unemployment and of pollution, congestion, crime, inadequate housing and overloaded facilities do exist, these are less a consequence of size *per se* than of intervening factors such as the form and structure of cities, speculation, poor city management, taxation policies and so on (Gilbert 1977). It is these latter policies that require modification and not enforced adjustments to size. Gilbert and Gugler (1982) add a note of caution about this. They argue that, while it may well be possible to design policies to cope with major urban problems, it may not be possible to implement them in developing countries where planning processes are highly politicized and development capital scarce. By inference, policies to curb increasing growth may in practice be the only alternative.

Some proponents of large cities deny that most developing countries have a problem of excessively large cities. Alonso (1975, p. 622), for example, argues that:

> . . . most developing countries seem to think they suffer from giantism of their principal cities. In some cases the worrisomely big

> city is quite small by comparison to urban areas in other countries, but looms large and is growing rapidly by comparison to other cities in the country. . . . The most common economic argument for calling this concentration excessive is the belief that per capita costs, particularly for infrastructure investment, rise after a certain urban size. However, there is no agreement as to the size at which this occurs, nor, for that matter, is there solid evidence that costs do in fact increase with urban size for a given level of service and facilities.

Central to the debate, therefore, is the issue of whether the cost of social overhead capital increases or decreases with city size and if it increases, whether this in fact cancels the gross income advantage. Evidence on this is limited. Mera's study of 47 developing countries indicates that, '. . . although the aggregate local government expenditure per capita rises after city size reaches a certain level, per capita income rises about four times faster than per capita local government expenditure for the same range, implying a marginal propensity to consume public services of some 25 per cent' (Mera 1973, p. 313). Even if it is assumed that quality of services does not vary with size, as Alonso (1975, p. 625) puts it: 'The engine of business is not economy but profit. That is to say, even if costs rise after a certain point, where productivity is rising faster (by reason of external economies of scale) big cities yield a greater net return per worker or inhabitant than smaller ones'.

At best, therefore, it must be concluded that to date the benefits of agglomeration relative to its diseconomies have not been costed adequately and the available evidence is inconclusive (Gilbert & Gugler 1982). The one thing that is widely accepted is that the concept of an 'optimum city size', which has been enthusiastically explored over the last three decades (Duncan 1957, Isard 1956, Shindman 1955, Klaassen 1965) is unachievable: as size increases, both costs and benefits change form. There is, in effect, a continually shifting point which makes the determination of 'optimal' impossible.

In policy terms, an essentially pragmatic compromise position can be found in the promotion of secondary cities (Rondinelli 1983). This approach has a number of objectives and underpinnings: it seeks to combat what is perceived to be the excessive size and primacy of the largest cities in many less developed countries; it rejects agropolitan and other primarily agricultural, bottom up, approaches on the grounds that they are impractical and that urbanization, industrialization and capitalization upon economies of agglomeration are essential to promote economic development; it seeks to stimulate *rural* development through increasing markets and sources of agricultural back-up; and it seeks to reduce spatial inequality through a more equitable distribution of national investment in forms that will benefit both the urban and the rural poor – in this regard, it is argued that 'the inextricable relationship between urban and rural poverty is seen most clearly in intermediate cities' (Rondinelli 1983, p. 39).

The secondary city strategy primarily stresses efforts to promote indigenous development in major cities and towns other than the primate city, as well as policies to slow down or even reduce the growth of the largest metropolitan area. Cities chosen usually function as high-order service centres for their surrounding rural areas. Strategies focus on measures to support rural development and to strengthen urban–rural linkages through the development of agro-industries, the production of agricultural inputs and the provision of social and utility infrastructure, including water supplies, sanitation, health, education and so on. They also include measures to stimulate small-scale industry and the informal sector. It is argued that only a small number of cities should be deliberately stimulated. A secondary city strategy is therefore one that concentrates on building on *strengths*. Advocates of the approach note that it should not be seen in isolation, but should be considered in the context of a national urban development strategy (Richardson 1981). In at least some versions of the strategy (Rondinelli 1983), restrictions to slow the growth of the largest cities are advocated.

While such an approach may well be appropriate in some contexts, it is by no means clear to these authors why this should always be so; there are several issues that are not adequately explained by proponents of the approach. The first set of issues relates to the question of the size and primacy of the largest cities. There are several dimensions to this.

First, not all the largest cities, whether or not they are pronouncedly primate, can be said to experience *problems* of size: certainly, cities the size of Harare in Zimbabwe, consisting of some 800 000 people in 1980, Luanda in Angola with approximately 750 000, or Nairobi, Kenya, with 900 000 in 1980 do not represent the same type of problem as Mexico City with 15 million people, qualitatively or quantitatively. Indeed, as proponents of the approach themselves acknowledge (Rondinelli 1983) the generative capacity of the larger cities, defined in terms of their ability to create income-generating opportunites for people is usually far greater than smaller urban centres or the rural areas. As great as problems of poverty in many cities in less developed countries are, standards of living in these areas are usually better than elsewhere. In countries where large numbers of people are being forced off the land through abject poverty to seek survival elsewhere, and where the *national* priority is the creation of jobs and survival chances, the degree to which governments can afford not to capitalize on every opportunity they have is questionable. In these situations, the largest cities represent a considerable resource and it may make sense relatively to concentrate urban investment on them, rather than actively attempting to restrict their growth.

Secondly, the approach does not always take adequate cognizance of processes of urban growth. Processes of urbanization in most less developed countries are very different from those in more developed parts of the world. In particular, rates of growth are faster and there is less balance between population growth and growth of urban economies. In almost all cases, population growth outstrips economic growth. Further, the largest cities characteristically have the fastest rates of growth, since migrants seek places of greatest opportunity – the largest cities. These factors often curtail the degree to which authorities can realistically channel investment to other urban centres on a significant scale. On one hand, they have little option politically but to invest where the people are. On the other hand, economic enterprises, responding to the availability of labour, capital and markets, continue to be attracted to the most propulsive points: the pattern is extremely difficult – and expensive – to break.

Thirdly, the secondary city approach is vague about how growth should be slowed in the largest cities. The danger is that *restrictive* measures are applied in order to encourage relocation or the diversion of new growth to secondary cities. International experience shows, however, that restrictive policies aimed at diverting *economic* growth may actually result in a national loss of jobs, if alternative locations are not genuinely viable and competitive. For example, it has been estimated that the introduction of the Physical Planning and Utilisation of Resources Act of 1967 in South Africa, which sought to divert industrial growth away from the major metropolitan areas towards outlying decentralization points by controlling the number of African labourers who could be employed by firms and by restricting the expansion of industrial land, resulted, in the first five years of its implementation, in the loss of 9.2 jobs for every 1 created in the alternative locations (Gottschalk 1977). Similarly, policies aimed at diverting *population* growth away from the largest cities inevitably discriminate against the poor. It is primarily the poorest people who are affected and these policies may lead to the growth of large pockets of poor people in locations where they are unable to benefit from opportunites and facilities generated by the more wealthy: the reinforcing quality of cities is dissipated.

The second set of issues relates to the definition and developmental impact of the secondary cities. Rondinelli (1983, pp. 48–9) defines secondary cities as 'urban places other than the largest city with population of 100 000 or more' and argues that these may have populations of 2 to 3 million or more. To argue that this range of city size constitutes, in terms of potential developmental impact, a single legitimate category is highly misleading.

First, the generative capacities, derived primarily though not exclusively, from economies of agglomeration of cities of, for example, 100 000 are entirely different from those of much larger cities and the propensity of policies to be successful is, therefore, at least in part related to size.

Secondly, the argument against primacy primarily hinges on the case that primate centres appropriate an excessive share of national investment and that primate systems limit the diffusion of innovation and other spread effects. In most developing countries, however, larger cities primarily occur in economically stronger regions: policies to promote these will do little to promote spread effects in those regions that require them most.

Thirdly, a primary justification for secondary city policies is that they promote agricultural development in the surrounding hinterland. It is true that increased market size may, in some instances, encourage increased agricultural productivity. However, in many cases there are other structural obstacles more serious than market size preventing significant increases in productivity, particularly among smaller farmers: issues relating to land tenure, access to credit, the flooding of markets by the output of large-scale estate agricultural enterprises, often directly or indirectly subsidized, to name but a few. The extent of this type of impact, therefore, will vary with context and requires careful analysis.

Conversely, there is no reason why policies to integrate urban markets and farming activity cannot be promoted *within* the primate cities. Properly, agriculture should be a significant urban structuring element in all large cities and with judicious urban management the number of rural producers who benefit from market proximity can be greatly increased.

There is no doubt that the concerns underpinning secondary city policies – the size and rate of growth of the largest cities and accelerating national and regional imbalance – are legitimate and real in many less developed countries. The possibilities inherent in such policies will, however, vary with context. It would appear that secondary city policies have their greatest prospects of success in contexts in which a system of significantly sized intermediate cities already exists and where policies are directed towards changing these from being 'exploitative' to 'developmental' (Rondinelli 1983, pp. 182–3): towards increasing the positive impact of these settlements on their surrounding hinterlands and through taking on problems of the urban poor more directly. Further, *how* the objectives of the policy are pursued is clearly critical. In most cases, restrictive policies on growth within the largest cities are dangerous and run a high risk of negative unintended effects that may well exacerbate the problems faced by the poor in the longer term. The stimulation of secondary cities then is best served by a positive system of incentives attached to them. However, international experience shows that, under capitalist systems in particular, it takes a long time to bring about significant adjustments to the space economy. The policies, therefore, should be seen as part of a long-term process of change. What can be achieved at any point in time requires careful contextual analysis.

Finally, secondary city approaches are not the only compromise ones advocated: some argue for a broader-based approach that includes an emphasis on both large cities and rural investment. As Mlia has pointed out: 'What seems to escape many people is that overall development and rural development are not necessarily incompatible or mutually exclusive' (Mlia 1975, p. 165). In a situation in which resources are scarce, it makes obvious sense to concentrate those activities which function most efficiently under conditions of concentration and to invest in the poorer regions in a manner which will bring most benefit to most people. Most significantly, it is necessary to identify goals clearly and to use the policy instruments most appropriate to achieving these: a single policy instrument can seldom be used to serve a multitude of disparate problems (Gilbert & Gugler 1982).

Core–periphery debates

A further set of debates relates to core–periphery conceptions of development and underdevelopment. Significantly, the core–periphery issue is different from issues relating to primacy, and the two do not necessarily go hand-in-hand. For example, while the core–periphery division is clearer in Zambia than in Tanzania, the reverse is true for primacy. In practice, however, core–periphery and primacy issues are often confused.

All the approaches discussed are underpinned, to a greater or lesser degree, by a core–periphery conception of development and underdevelopment. Interpretations of the consequences of core–periphery relationships, however, vary substantially. The growth approaches – both those deriving from the older sector and export-base theories and the newer accelerated development approaches – regard the core–periphery relationship as positive, though it is perhaps less important as a development tool in the accelerated growth approach than in earlier growth theories. In terms of *spatial* interpretations of sector theory, economic cores are often equated with the largest urban centres (Friedmann 1972). These *urban* cores provide the initial generative impulses of development which are then diffused downwards and outwards into the modernizing periphery. The diffusion process occurs via the settlement system, from larger to smaller centres and thus requires the development over time of an integrated rank-size settlement hierarchy. The accelerated growth approach accepts the concept of diffusion, but does not pay much attention to the rôle of settlements in the diffusion process. In both cases, it is assumed that the optimal hierarchy of settlements will develop more or less automatically: core–periphery imbalance is a self-correcting condition if the right economic conditions prevail and, therefore, so is primacy. Both sets of theory, however, have tended, to a greater or lesser degree, to argue for policies to combat the initial imbalance. In the case of sector theory, it is believed that the manipulation of the settlement system in the direction of a more balanced hierarchy will have a positive impact on economic growth. In both cases, there is a concern with curbing 'urbanization without industrialization' and hence urban unemployment and its expected political consequences.

All the other approaches, to a greater or lesser degree, see the relationship between core and periphery as negative and central to the problem of underdevelopment. The core is viewed as a locus of political and economic power which systematically absorbs, to its own ends, the most productive resources of the periphery. All have, as a specific objective, a reduction in rural–urban migration and all aim to develop the periphery against the core. Further, the dynamic in the relationship between core and periphery is viewed as tending not towards equilibrium but towards increasing polarization.

This conception of the core–periphery relationship, in so far as it takes spatial form, can be criticized both on theoretical and on empirical grounds. At a theoretical level, the central concept that a core (an area) exploits a periphery (another area) is seriously questionable. The idea derives partially from a crudification of Frank's (1971) concept that one nation exploits another. At a national level, Frank's concept as an explanation of underdevelopment has generally been disputed because of its failure to look at internal factors: most particularly, the rôle of the state and class struggle, the specific way in which capitalism has developed in a nation and the way in which these factors interact with the particular form of insertion of the nation into the world capitalist economy. It is argued that Frank's version of underdevelopment is mechanistic at best and incorrect at worst. There is a clear distinction between spatially uneven development and underdevelopment, which relates to the welfare of people (Gilbert & Gugler 1982). Whatever the validity of any element of the concept at a national–international level, however, it is not possible simply to transplant concepts to a lower level of spatial aggregation (Massey 1978). There are empirical differences between a nation and a region: taxation, control over monetary and trade and customs policies, to name but a few. The rôle of the state, and class relations in it, are usually different at regional and national levels. Further, space is not an *abstract* entity governed by 'laws' which occur at any scale: rather, 'we are dealing with *social* divisions of territory and socially different types of territorial division' (Anderson 1978, p. 15).

At a more specific level, it is questionable whether the core–periphery idea is adequate analytically. The core–periphery concept is widely used to describe very different *types* of processes and different forms of uneven development: for example, relations between urban and rural areas in different contexts; relations between areas that are capitalist and those that are pre-capitalist; relations between areas that are more and less wealthy; processes of urbanization and primitive accumulation; conditions of primacy; simultaneous tendencies of capital to concentrate spatially and to diffuse its products; relations between branches of industries, branches of firms, and so on. It is increasingly being argued that these forms of uneven development – and, significantly, 'core–periphery' is not the *only* form of uneven development – must be understood as instances of the way in which capitalist development produces configurations in accordance with its laws of motion at a point in time and in interaction with the socially defined territory upon which it operates. The study of uneven development therefore requires identification of precise, contextual elements and not simply broad references to 'core–periphery'.

The 'core–periphery' interpretation underpinning agropolitan development, selective territorial closure and, to a small extent, some versions of basic needs, is based largely on the idea that the forces of 'modernization' break down 'traditionalism', and this results in the underdevelopment of 'traditional' areas. This, as reflected in the work of Frank, is essentially a reversal of the dualist hypothesis. However, it has since been shown that the forces of modernization in developing countries have not always dissolved traditional or pre-capitalist modes of production. Rather a complex process of 'conservation-dissolution' has occurred, the nature of which has depended on changing conditions of colonialism and international capitalism and on processes of class conflict (Laclau 1971). Further, these processes of 'conservation-dissolution' tend to build new inequalities and to exacerbate *existing inequalities*: this is a far cry from the idealized and equal rural conditions depicted, for example, in Friedmann's work since 1976.

Agropolitan development, selective territorial closure and, to some degree, basic needs, seek either to protect peripheries from core-directed disintegration, or to set in motion a process of 'conservation' where the disintegration of the 'traditional' has occurred. Arguably, there are few 'peripheries' that are entirely unexposed to the capitalist mode of production and, for the rest, it is possible that distorting processes are too deep to be 'cured' simply by regional autarky. For example, in many contexts, local power structures have been transformed and distorted by a variety of external influences including, in some cases, colonialism. The resultant structures are usually by definition undemocratic and self-serving. Further, where peasant production has generally disintegrated, or, at any rate, been highly marginalized, it is debatable whether it is always possible or desirable to attempt to re-create peasant production. In some contexts, for example, the cost in financial terms may be too high. Similarly, skills may have been lost and a demand for industrial products engendered. This demand is unlikely to disappear simply because planners or decision makers wish it were different. Where reasonably developed agricultural and industrial sectors exist at a *national* level – that is, particularly where industrial production at regional level is not necessary to fulfil national demand – regional closure may simply stunt development at a national level and result in reduced efficiency and growth overall. If, for example, effective barriers are erected against national industries to protect small-scale local production, the overall effect, if the policy is 'successful', will be to weaken large-scale industries for the dubious benefit of developing less efficient industries in peripheral areas. It is likely that these industries will both pay workers less than larger 'core' industries and produce more expensive goods. In these circumstances, the welfare objectives of the approaches are lost through locational over-determination.

Criticisms have been levelled not only at the core–periphery concept: the anti-urban bias which derives from it has also been attacked (Sandbrook 1982). It is argued that the analysis upon which the anti-urban position is based is faulty on two counts. On one hand, the anti-urban view assumes – incorrectly – that poverty is essentially a rural problem and that the urban poor are transitory rural dwellers. However, they are often working members of the informal sector and, in most cases, are committed to urban life on a long-term basis. On the other hand, it assumes that rural élites generally possess extensive urban connections. As Sandbrook (1982, p. 19) points out, 'This is a convenient view. But it does not often accord with reality.' Further, the tendency to juxtapose urban and rural overestimates the degree of equality in rural areas. It ignores the fact that many factors, including colonial development and, more recently, attempts to commercialize agriculture, have resulted in many places in a process of rural differentiation. From this point of view, it is not possible to objectify rural areas and to argue that 'urban' exploits 'rural'. In fact, there is often significant absolute poverty in urban as well as in rural areas, and, despite the overall urban bias in services (Lipton 1977), the poorest urban dwellers seldom have access to these. The bias in service provision, too, is often a *class* bias: for example, education institutions in low income areas are scarce and of poor quality and are often no better than those in rural areas; health care for the urban poor is often inadequate, distant from low income areas and usually too expensive to use (Sandbrook 1982).

Primacy debates

A third debate, closely interrelated to the others but conceptually quite different from them, hinges on whether or not a condition of primacy, where one or a limited number of cities dominates the rest of the settlement system, is developmentally positive or negative. Two issues are central to the debate. The first is whether or not there is a correlation between primacy and economic growth. The second relates to primacy and its impact on innovation diffusion.

In relation to the first issue, there have been several studies that have attempted to correlate levels of economic development with primacy. Mera (1973), relating increasing rates of primacy to increasing rates of growth of gross national product (GNP) across 47 developing countries, over a period of seven years, found that those countries for which a change in primacy was less than 1 per cent in general had a much lower growth of GNP per capita. Those countries for which change in primacy was over four percentage points had a growth rate of GNP much higher than any other nation. Berry's attempts to correlate levels of development with primacy led him to conclude that different city size distributions are in no way related to the relative economic development of countries (Berry 1961), a conclusion disputed by some (Stöhr 1974). El Shakhs (1965), on the other hand, argued that the relationship was a non-linear one, with 'low primacy at the early and advanced stage of development and a peak at the middle range' (El Shakhs 1965, p. 47). Even where a correlation between primacy and development is claimed, it begs the question of which is cause and which effect. Further, it is conceded that an intervening variable, 'economies of agglomeration', which are largely a function of city *size*, must be considered. Thus, higher labour productivity in larger cities may be traced to factors such as greater specialization in services, better quality schooling and on-the-job training and greater density of social overhead capital and economic overhead capital. At best, therefore, available evidence appears inconclusive. Clearly, primacy is not *inherently* negative in terms of economic growth: it appears that conditions vary with context and that the specifics of each case require analysis.

The second, and primary, issue hinges on the relationship between primacy and *regional* development, as opposed to national economic growth: particularly the rôle of innovation–diffusion processes. This will be discussed further in the following section.

Debates relating to the development of a functional hierarchy of settlements

Different approaches to development have different implications for the functional and thus, usually, the size hierarchy of settlements. The most precise articulation of the general form and rôle of such a hierarchy in promoting development can be found in sector theory. In this conception, primacy tends to emerge in early stages of economic growth and the hierarchy adjusts from a primate to a log-normal, rank-size distribution as development occurs. The development of this form of hierarchy is not simply a consequence of processes of development but is vital in terms of the *promotion* of modernization: modernizing impulses are transmitted downwards through the settlement hierarchy and outwards into the periphery, and the different levels of the hierarchy play specific rôles in the promotion of development.

The accelerated growth approach *implicitly* accepts the conceptualization of the diffusion process. In terms of settlement policy, it focuses on medium-sized and smaller settlements in the strongest regions. By implication, the approach does not necessarily seek to promote a log-normal, rank-size distribution: its application would, over time, result in an autonomous, market-determined hierarchy which would, arguably, be skewed in favour of the top half of the curve: it would result in the development of a limited number of growing large and medium-sized towns.

Redistribution through growth approaches tends to advocate policy measures to develop the bottom end of the settlement hierarchy in order to aid the modernization of agriculture. Authors such as Johnson (1970) and Rondinelli and Ruddle (1978), whose work, like redistribution through growth proponents, emphasizes the modernization of agriculture and focuses on redistributionary forms of growth, stress the need for a fully-developed, integrated functional hierarchy. For Johnson (1970), the lack of a nested central-place hierarchy leads to conditions of monopsony in local markets, a factor which in turn keeps prices low and hence reduces the incentive to produce. A functional hierarchy is necessary to allow rural producers access to competitive markets and to encourage money exchange and the buying of inputs for agriculture. Spatial structure, therefore, is seen as a possible impediment or an encouragement to the development of a progressive rural social structure. Rondinelli and Ruddle's work proceeds on similar lines. In terms of a rural modernization approach, therefore, a hierarchy of urban centres is seen as a means of providing a rational and competitive form of spatial organization for marketing, dispersed industry and informal sector activities; for decentralized administration and rural support services; and for social and utility services. For Rondinelli and Ruddle (1978) support services tend to occur in a hierarchical form: this underpins the need for an entire functional hierarchy and not simply for a single level of small settlements.

Basic needs approaches do not explicitly advocate the promotion of an integrated hierarchy of settlements although they, too, call for the existence of a regular distribution of small centres, controlling sufficient threshold support to allow for the efficient and viable provision of social and utility services. As in redistribution through growth approaches, however, spatial planners, such as Mayer (1979), have advocated an integrated hierarchy of settlements to accommodate the different ranges and thresholds of various levels of services and to concentrate levels of services with similar ranges and thresholds.

The agropolitan development and territorial closure approaches are, in many respects, mirror images of sector theory. They envisage the emergence of an articulated settlement hierarchy which provides the framework for equal territorial development and which is integral to the organizational framework of society: each level of the settlement hierarchy is associated with a certain level of 'bottom up' control as well as with a certain level of functional organization. Significantly, however, this system is developed from the bottom up rather than from the top down. The rôle of settlements is to provide support systems for local agricultural production and settlements are thus supported 'from below' by their immediate hinterlands. Since urban centres are primarily orientated inwards to their hinterlands, the growth of large cities would be curtailed. The slope of the rank-size curve would thus be flatter than in the case of sector theory. The dynamic of the system is again reflected in a diffusion process but, in this case, developmentally positive impulses, as well as economic surpluses, are perceived to 'trickle up' to successively higher levels rather than to 'trickle down' to lower levels. The mechanics of the diffusion process are not identified. The *spatial* distribution of small centres in all approaches is determined by considerations of agglomeration economies, adequate thresholds and reasonable access. The primary *principles* of central-place theory, as opposed to the special form of the Christaller model, therefore, hold ubiquitously in determining either the location of small centres or priorities in terms of settlement stimulation.

The major policy debate relating to the settlement hierarchy hinges upon whether it is desirable, or possible, to promote a fully developed and integrated urban hierarchy in developing countries. Proponents of an integrated hierarchy argue in terms of the utility of the idea and hinge their arguments on four main points. The first is that the development of a fully fledged urban hierarchy can be used to provide a spatial framework for different types of development policy. Hence, the upper tiers of the hierarchy can be used to stimulate large-scale industrial development, including growth poles. Lower tiers can be used to aid rural development and small-scale industrial development by providing a locus for marketing, agricultural support activites and services. Intermediate tiers can be used to provide a location for higher-order support services and for industry that processes agricultural products. If properly used, it is argued, rural and urban activities can be integrated by way of the hierarchy with, for example, growth poles developed on the basis of industrial links up and down the hierarchy. The policies proposed for Nigeria, India and Brazil by Mabogunje (1978), Misra and Sundaram (1978) and Babarovic (1978) exemplify ideas of this sort.

The second argument is that an integrated settlement hierarchy is necessary to break down colonial or neo-colonial enclave development (El Shakhs 1974, Rondinelli & Ruddle 1978). Arguments of this sort were applied, for example, in the cases of Tanzania (Piöro 1972) and Kenya (Soja & Weaver 1976). The third argument as, for example, Rondinelli and Ruddle (1978) contend, is that the lack of a functional hierarchy results in a failure to stimulate agricultural development, as a 'population scattered in small hamlets and villages does not permit large enough concentrations to form regular institutional markets for higher agricultural productivity. There is little reason to save and invest; specialization and divisions of labour do not occur; and opportunities for expansion and non-agricultural employment are few' (Rondinelli & Ruddle 1978, p. 55). Further, without it, it is impossible to promote an efficient support system for rural development, adequate services or administration decentralization in a spatially logical way.

Finally, it is argued by theorists operating from the premises of the modernization paradigm that an integrated functional hierarchy is critical to the diffusion of innovations and this is central to the process of modernization.

Arguments against attempting to establish an integrated urban hierarchy centre around the inappropriateness of the policy to conditions in developing countries. One argument is that particular forms of settlement systems are associated with specific experiences of development (Harvey 1973). The theory of a functional hierarchy of settlements and its utility in development is based on the experience of developed countries: however, it is widely argued (see, for example, Frank 1971, Gilbert & Gugler 1982, Leys 1975) that the development processes occurring in developing countries are very different from the experiences of more developed countries. Consequently, it is unlikely that a similar form of functional hierarchy will occur and its utility to development planning in developing countries is therefore questionable. A second argument relates particularly to urban-industrial functions and holds that it is simply too expensive to spread these functions widely into the periphery.

Perhaps the most telling argument, however, is that the unproblematic use of these approaches in some contexts represents a misunderstanding of the nature of rural accessibility and demand. The implicit assumptions underpinning the approach are that there is a relatively even spatial distribution of demand and that patterns of accessibility are reasonably equitable. In many less developed areas, neither of these conditions holds. There are several dimensions to this.

One is that contextual factors – such as topography, climate, levels of poverty and so on – profoundly affect rural accessibility and the static nature of these rigidly hierarchical approaches renders them incapable of accommodating these realities. This relates to rural producers and consumers alike.

Many rural producers, in both the small-scale manufacturing and farming sectors, are effectively immobilized through poverty. They do not seek to sell products from the theoretically most accessible points or urban centres, since they have no access to these. They attempt to optimize location only within the very tight constraints of where they are able to reach on foot (Preston-Whyte & Nene 1984). It is precisely this fact that underpins the high degree of monopsony in many rural areas (Johnson 1970). In many cases, there is a considerable monopoly control over transport (Carapetis *et al*. 1984). Many small farmers cannot afford access to vehicular transport when they have surpluses to sell. For want of transport, they are locked into local markets which, in very poor areas, are by definition small, weakly developed and easily flooded. Usually, the only people who do have access to transport are the middlemen, such as rural traders (Lienbach 1983, McCall 1977). This control over movement, often reinforced by control over storage facilities, creates a situation of monopsony: farmers have to sell, but middlemen do not have to buy and accordingly they exert considerable control over prices. Effectively, therefore, the marketing system is 'dendritic' (Johnson 1970): there is little integration of local markets. These conditions are entrenched in many areas through 'debt servitude': traders grant credit to farmers on condition that they control the distribution of the surplus. In these situations, the primary benefits of road improvement programmes and other forms of rural investment are reduced costs for middlemen and traders and few of these benefits are passed back to the rural consumer. An evaluation of a World Bank-financed rural road project in Iran (Mabogunje 1980, p. 292) for example, revealed that transportation costs for agricultural produce dropped by 20 per cent but prices paid to farmers remained unchanged. The monopoly position of transporters and middlemen allowed a difference of 25–50 per cent between the farm and the Tehran prices, whereas the transport costs accounted for only 5–10 per cent of the margin. Similar results were found in an evaluation of a feeder road project in the Yemen.

Rural consumers, too, are differentially affected by a range of factors affecting accessibility, with the net result that there is no clear, direct correlation between physical distance and the range of goods, a service or a rural centre. One dimension of this is the fact that the opportunity cost of time in rural areas varies widely in response to a broad range of factors. Thus, for example, in one case experienced in the Transkei Bantustan of South Africa, two girls had walked two hours to spend 70 cents at the local store, partly because there was little else to do at that time of year and partly because the store constituted a social focus and a meeting point. Conversely, rural studies in India (for example, Willbanks 1972) reveal a situation of much tighter time constraints, because of higher levels of productivity and more sophisticated agricultural programmes, less seasonal activity and so on.

A second dimension is that factors affecting accessibility often reinforce high degrees of monopoly in rural areas. Many people and communities are locked into the sphere of influence of single traders and are unable to break out of it, either because alternative outlets are located considerable distances away and people cannot afford the associated time and monetary costs to overcome the friction of distance, or because roads are so bad that they are effectively entrapped. Often, therefore, prices in rural trading outlets are considerably higher than those found in the towns.

A third dimension is that in many least developed areas, particularly those in Africa, incomes are too low and the size of market too small to support a full range of functional levels in the hierarchy. For example, Funnell (1973) found that, as a result of low incomes in the Teso region of Uganda, there was little functional differentiation in commercial activities between medium- and low-order settlements. The main difference was in the size of stock carried by shops in the different centres. In situations like this, consumer decisions are not only affected by issues relating to distance: people do not inevitably visit the closest facilities which offer the goods or service required, but another set of factors, such as greater choice and cheaper prices, or the possibility of combining a number of functions, including social functions, into a single trip, comes into play. In complex situations like this, artificial attempts to impose a functional settlement hierarchy where none exists will result in the wasteful use of limited public resources and the duplication of the same functions at different levels of the hierarchy.

These kinds of problems are magnified in poor areas where population densities are low. Use of rigid hierarchical approaches in cases like this represents a compromise: although service centres are located at some theoretically acceptable distance from rural dwellers, in practice this may be too far for people to travel for frequently required services. Effectively, therefore, lower and middle ranges of the hierarchy compete with each other, dissipate resources and incomes and, overall, lead to extremely low levels of accessibility.

A further criticism that can be levelled at the unthinking use of the approach is that in many contexts its application represents a misunderstanding of the dynamics of regional spatial order and growth. Different services and different economic activities have different propensities to be mobile and different time rhythms of usage: if a fixed service is required daily but is located outside of the range of daily use – and the definition of this will vary with local movement technology – it will not be used and the need will remain unsatisfied. In this case, the settlement is as inaccessible as other centres that are located much further away. Further, different services have different thresholds for viable provision. Obviously, higher-order services require larger thresholds and therefore must serve larger areas. Faced with these factors, it is immediately apparent that a rigid policy of artificially concentrating services in service centres cannot work in all cases. The difficulty lies in defining the order of settlement in which concentration occurs. If the choice is upon an intermediate order of centre, people do not have easy access to the lowest-order services, which are often the most important. However, if the lowest-order is chosen, intermediate-order services often cannot be provided viably.

Finally, the point-related investment pattern inherent in the hierarchical approach may lead to high levels of accessibility, *provided that a full range of centres and a fully developed hierarchy exist.* If they do not and, as has been argued, the possibility of creating a full range of centres in most peripheral areas is remote, large areas remain entirely unserved. The approach, then, is static and simplistic: acceptance of it ignores the complexities of growth over time. This problem of inaccessibility is compounded by the transport-related implications of the approach, which calls for an interlinked radial transport system with routes converging on, and thus reinforcing the centrality of, these centres. If gaps exist in the settlement hierarchy, distances between radials become excessive further away from existing centres. Large areas are therefore effectively unserved by the transport network.

Recognition of these difficulties does not imply a blanket rejection of hierarchical approaches or of rural service centres. Obviously, there are advantages in agglomerating services – particularly gaining economies of agglomeration and increasing convenience, by enabling the consumer to satisfy a number of needs simultaneously – where this is appropriate and where the advantages of such an approach outweigh the disadvantages. In the final analysis, however, the problem of access to opportunities and facilities in poorer rural areas is much more one of demand than of supply. As a general proposition, the possibilities of using an integrated hierarchical settlement approach successfully are highest in areas that already contain an established settlement hierarchy and where rural densities and incomes – and thus thresholds – are relatively higher. Clearly, however, the approach cannot be applied rigidly or ubiquitously: the determination of the most appropriate form of spatial order requires careful contextual analysis.

Debates relating to small urban settlements

Significantly, all the approaches to development which have emerged in the 1970s and 1980s have stressed the generation and stimulation of small urban centres in rural areas. An important difference in emphasis, however, exists between various approaches regarding the rôle of these urban centres in promoting rural development.

Sector theory, accelerated growth and, to a lesser degree, the redistribution through growth approach emphasize the concept of 'rural modernization': that is, the idea that the aim of rural development is to commercialize the peasantry. It is believed that rural productivity will be raised if urban centres, and thus markets, are made accessible and that peasants can be induced to produce more by exposure to the goods and services available in urban centres, since these will affect motivation to earn more. This urban–rural exchange function is thus best promoted by dispersed urban development. In addition, rural modernizers argue that small urban centres can absorb population from the land in the long run and this again aids the development of agriculture (World Bank 1981). In some versions of the redistribution through growth approach, rural modernization is coupled specifically with the stimulation of *rural* industrialization and by the spread of a range of services delivered through the urban centres. By contrast, agropolitan development and, to a lesser extent, basic needs, attempt to move away from the association of 'urban' with 'modernization'. They argue for a blurring of a functional distinction between urban and rural, particularly by promoting rural industrialization and hold that small urban centres should provide an essential life support system to local development. In these approaches, urban centres can *support* rural areas by providing support services for agriculture, a locus for rural industry, a means of absorbing that surplus rural population who wish to remain in rural areas and a physical structure for co-ordinating local political, social and economic development. Significantly, however, a basic needs approach is not wholly dependent on the existence of small urban centres. Much of the infrastructure that would be provided for an area under a basic needs investment programme, such as electricity, water and sanitation does not necessarily take on urban form.

Despite this essential difference of emphasis, all approaches perceive the small urban centres as providing a number of functions for their surrounding rural hinterlands. The stress laid upon different functions, however, varies with approach.

The functions of small urban centres

Marketing The marketing function is central to the accelerated growth and redistribution through growth approaches in particular. Basically, they argue that efficient marketing is essential for commercialization but in many rural areas it is impeded by poorly maintained roads and the absence of a functional hierarchy of settlements. Both these problems make the collection of produce difficult and the time delays and poor roads cause goods to spoil. Further, low volumes of production and a lack of product standardization often make it difficult to organize a state marketing system. There is therefore a need for centres that can provide the thresholds necessary to consume surplus products produced on an irregular basis. If such a settlement network does not exist, competitive markets do not arise, there is a lack of information about prices and there is insufficient stability of demand to encourage sustained improvement in productivity and output. In both approaches, therefore, small settlements and the transportation network are orientated outwards to larger centres and larger markets.

In the basic needs approach and, to a lesser extent, in the agropolitan and territorial closure approaches, marketing as a condition for growth is not stressed. The emphasis in these approaches is upon local – interpersonal – marketing of foodstuffs. Small centres are still important as a locus of efficient exchange but, in the definition of efficiency, the approaches tend to stress consumers as opposed to producers. Badly maintained roads and inaccessible or very distant market centres are still regarded as negative, but primarily because they impose higher costs on consumer goods and may make rural living more expensive, and certainly much more inconvenient, than urban living. In the agropolitan approach, the interconnection between different sizes of markets, and thus of marketing centres, is not so important: urban centres and transport networks are internally orientated towards their immediate hinterlands, rather than externally orientated towards larger markets.

Industry All the approaches stress, to a greater or lesser degree, the importance of small urban settlements as a locus for industrial location. Generally, these industrial activities are of two kinds: agro-industries using raw materials from the rural hinterland and small, often informal, manufacturing industries producing cheap, locally used, consumer goods from local resources. There are three main reasons for the emphasis on small urban centres in this regard. First, even small-scale concentrations allow for the emergence of localization economies. In particular, services such as electricity, clean water, suitable premises and finance can be provided more cheaply and efficiently. Secondly, concentration allows for a more efficient system of marketing. Thirdly, the stimulation of small industries requires organizational and financial inputs. The concentration of small-scale industries in small centres allows for the rationalization of organizational assistance and for a focusing of developmental efforts. Further, it is anticipated that the presence of these industries in proximate urban centres will provide, at least in part, a means of absorbing population from agriculture (Rondinelli & Ruddle 1978).

The emphasis on this type of small-scale industrial activity is strongest in the redistribution through growth, radical basic needs and agropolitan development approaches, where it is seen as a means of transforming the structure of production and consumption. It is most marginal in the accelerated development approach, which basically argues that industrial activity in rural areas will largely follow rural modernization and improvements in rural incomes and that emerging industries will locate at places which, in terms of the logic of the market, are the most efficient.

Innovation–diffusion In the rural modernization approaches characteristic of redistribution through growth and accelerated development, small urban centres are seen as one link in a chain of innovation-diffusion leading towards national economic integration. This perception stands in direct contrast to the internally orientated regionalism of the 'development from below' approaches, which view the urban centres and their hinterlands in a more autonomous way.

The 'capture' of income leakage In most of the approaches, small urban centres are intended to play a part in capturing the benefits of rural development programmes and in preventing income leakage from local regions. Most commonly, income leakage occurs as a result of spending on consumer goods in outlets outside of the region, through the transfer of profits out of the region by multibranch stores and through the importation of raw materials by manufacturers and the transfer of industrial profits out of the region. The degree to which this leakage factor can be controlled affects, in part at least, the degree to which the 'multiplier' factor occurs within the region. The rôle of small towns in this case is to provide a location for the establishment of small commercial and industrial establishments which will have strong linkages within the region (Whitsun Foundation 1980).

Service provision In most of the approaches discussed, the rôle of small urban centres in the efficient and viable provision of essential services is extremely important. Subtle distinctions relate, however, to the way in which service delivery is perceived.

The equitable and immediate delivery of basic *social* and *utility* services essential for the satisfaction of basic needs – health, education, potable water, sanitation, housing, communications and commerce – is central to the basic needs and agropolitan approaches to development. In the redistribution through growth approach, this kind of service provision is perceived to follow increases in rural and small-scale industrial productivity and output. Agricultural back-up services – extension, research, repair and maintenance, and financial services – are, however, considered as central to improving agricultural productivity. In this sense, therefore, service provision is central to the concepts of innovation-diffusion and rural modernization embodied in the redistribution through growth approach. The provision of agricultural back-up services is also critical in the agropolitan development and, to some extent, in radical basic needs approaches, although here the *form* and *content* of the services is somewhat different. The provision of a rational location for mass-based organization and decentralized administrative structures is stressed in all three approaches, although in slightly different forms. Redistribution through growth stresses the practical aspects of these functions, while basic needs and agropolitan development emphasize the rôle of local mass-based organizations in mobilizing rural development (Friedmann & Weaver 1979, Friedmann 1980, Stöhr 1981). Radical basic needs and agropolitan development proponents also stress this function since the poor gain greater control over the decision-making process. In accelerated growth approaches, little emphasis is placed upon state intervention in the equitable delivery of services. Intervention is at best focused on 'productive' services (for example, agricultural extension) and upon the strongest regions. It is assumed that the market will deliver social and utility services at the most efficient places, according to the dictates of effective demand.

All the 'bottom up' approaches – basic needs, agropolitan development and, to a lesser extent, redistribution through growth – lay stress upon appropriate *forms* of service delivery: *how* and in *what form* services are provided are considered to be as, or more, important than the presence of the service itself. The connection between service delivery and urban centres lies, first, in the need to create adequate thresholds to provide services viably and, secondly, to minimize the friction of distance for rural consumers. The concepts of range and threshold, therefore, which have long been central to central-place theory, remain in the forefront of small urban centre planning. Significantly, while the objective of promoting a settlement system that maximizes the effective access of people to the widest possible range of services has ubiquitously been an important part of rural settlement planning, not all *forms* of that planning have focused on small urban centres. In practice, four main forms of planning of small settlements dominated in the developing countries in the 1970s. These fall into two broad categories: strategies to transform the pattern of land occupancy – nodal rural development strategies and the creation of villages; and strategies to restructure rural areas as a whole – rural service centres and the use of mobile services. A third category of rural settlement planning, although more common in developed than in developing countries, has been advocated in some developing countries and relates to attempts to prevent the decline of infrastructure in small urban centres. The strategy most commonly advanced is the promotion of 'holding' centres – see below, p. 155.

Forms of small settlement planning

Nodal settlement schemes Nodal settlement schemes involve the concentration of investment over a limited area and on a limited number of people. They frequently take the form of irrigation schemes and are often run on government estates. Examples of these can be found in most developing countries. In Malawi, tobacco, sugar and tea estates have been set up by the government in conjunction with foreign companies (Christiansen & Kydd 1983) in the belief that economic development can be based on the expansion of large-scale agriculture. In Zimbabwe, the pre-independence government set up 'rural growth points' based on irrigation schemes, largely to extend state control into the rural areas (Riddell 1978). The schemes are normally capital-intensive and are aimed at encouraging the commercialization of agriculture. They usually involve the provision of intensive training and other support systems to the more successful smallholders, in the belief that the impact of organized markets, services and, particularly, extension programmes, will spread to the surrounding region. The idea, generally, is one spatial form of the capital-intensive approaches to agricultural development which dominated regional planning in the 1960s. More recently, however, the concept, if not the form, of these earlier approaches has been included in accelerated growth programmes. In practice, the track record of these programmes has been poor. Benefits in general have failed to 'trickle down': individual input packages have reached few people and have proved too costly to replicate on a widespread basis and the form of extension services offered has often been inappropriate to conditions prevailing outside the nodal scheme itself. Further, schemes have proved very expensive to maintain: an initial pattern of concentrated investment has led to an enforced continuation of the pattern over time (Riddell 1978, Dumont 1979, Dewer *et al*. 1982, 1984).

The creation of villages A common policy approach has been the removal of families occupying the land in a dispersed pattern in rural areas, and their relocation in a series of concentrated planned villages. Such an approach has usually had four main aims: to provide basic, first level services, such as clean water, primary education and primary health care; to modernize agriculture by consolidating land holdings into units large enough for the efficient use of agricultural machinery – where appropriate – or to stimulate the emergence of co-operative and collective forms of production; to promote optimum land use by conserving the richest land for cropping, drier lands for grazing and worst agricultural land for residence; and lastly, to improve farming practices and, in particular, to combat erosion and to promote soil conservation.

The determination of the size of villages has been dominated by the concept of minimum thresholds, particularly for education and water provision: a figure of 250 families is often held to be the minimum effective size, though villages are often considerably larger. The spatial *distribution* of villages is determined by the need for concentrated populations to reach their lands, using locally available transport technology, on a daily basis. In some cases, these policies have been introduced for political reasons, a motive which has been more overt in some countries than in others. In certain, particularly socialist, countries (for example, China) the policies have been used in an attempt to combat the power of wealthy landowners and in order to establish a spatial framework for political organization. In the case of the Ujaama policy in Tanzania, which is probably the best documented case of 'villagization' (see Hyden 1980, Von Freyhold 1979, Raikes 1975) the intention of the programme was to transform the countryside into a set of villages which would aid the collectivization of agricultural production, the provision of shelter, mutual aid and popular participation in decision-making matters affecting the rural areas, servicing and marketing. Village development would provide a vehicle for the diversification of the rural economy and this, together with improved standards of living, would arrest urbanization (Mascarenhas & Claeson 1972). In other cases, such as the 'betterment schemes' used in the South African Bantustans, it has been used as a means of social control and as a way of extracting labour from the countryside (see Yawitch 1982).

The track records of village-creating programmes have been variable and have generally reflected the appropriateness of the plans themselves and the way in which those plans have been implemented. Common criticisms levelled at these programmes include their exorbitant cost, the massive social upheaval resulting from relocation, the inefficiency of the daily journey-to-work patterns imposed by the schemes and the inappropriateness of many schemes to local conditions. In some cases, village creation has not even been accompanied by the provision of necessary services. The Tanzanian programme, particularly, suffered from bureaucratic initiatives that failed to respond sensitively to local conditions and failed to recognize that in certain areas, village creation was an inappropriate strategy for both social and agricultural development (Raikes 1975).

Rural service centres The rural service centre concept represents an attempt to use settlement planning systematically to restructure rural areas and, in particular, to provide a spatial framework for marketing and for social service delivery. It is heavily informed by central-place theory and, in particular, the Christallerian forms of that theory. In this conception, the most important vehicles for service delivery are small 'service centres' which occur frequently in order to maximize access: despite enormous contextual differences which profoundly distort the concept of 'accessibility' – topography, population distribution, income distribution and so on – it is almost ubiquitously proposed that these centres should be evenly distributed and serve a radius of between 5 and 30 kilometres. Most commonly, although not always, the rural service centre is seen as the lowest significant level of a rural settlement hierarchy and may often lie at a lower level than small urban centres. Where villages exist or are created, they are usually seen as one tier of the service hierarchy. In their absence, the lowest level of settlement usually has limited functions which would otherwise occur in the village, for example, primary education, a shop, possibly primary health care and so on. Different forms of hierarchy have been suggested in different contexts. Rondinelli and Ruddle (1978), for example, advocate a three-tier structure: village service centres, to focus rural activities and to raise the quality of rural life; market towns, to stimulate the commercialization of agriculture; and an intermediate city, to absorb migrants and co-ordinate development impulses from 'above' and 'below'. By contrast, Friedmann's agropolitan concept for Mozambique (Friedmann 1980) relies mainly on a lower-order 'silent centre' which provides basic daily services – at least a shop with expanded activites – and a more developed District Service Centre at a higher level than Rondinelli and Ruddle's Village Service Centre. The District Service Centre, unlike the Village Service Centre, is 'a major locus of development activities' (Friedmann 1980, p. 106), and includes the full range of services for economic, social and political life.

Criticisms of the rural service centre concept have been advanced on a number of grounds. One is that the concept is static and inflexible and thus cannot respond to growth or change. For example, if incomes in rural areas are raised, people may bypass the service centre in order to use higher-order centres. Further, the concept relies on an assumption of a spatially even distribution of demand and production costs over a uniform plain: criteria that are rarely met (Reynolds 1981). The approach is costly to implement and its daily running makes heavy inroads on scarce managerial resources. Another criticism is that while rural service centres are theoretically located at some 'acceptable distance' from users, in practice they are too far for frequent trips and the services provided are not specialized enough to warrant major trips. In parts of India, for example, studies show that poor farmers travel frequently over very short distances to achieve single purposes, but travel infrequently – four to five times a year at most – to central places (Reynolds 1981). The concept can also be questioned on the grounds of its vagueness: seldom are attempts made in different contexts to determine the ranges of different services and, it can be argued, this type of analysis may well invalidate the concept entirely. Finally, it has been argued that in certain rural areas, blockages to development are not caused by a lack of *supply* – of basic or necessary services and production facilities – but rather by insufficient *demand* caused by factors such as low prices for peasant output (Christiansen & Kydd 1983). In this case, the removal of 'supply constraints' will not result in self-sustaining growth. This point is echoed by Gilbert and Gugler (1982), who comment that blockages to rural development are usually complex and varied: the simple creation of rural service centres, on their own, cannot rectify the manifold problems of the rural areas. While not constituting an argument against the establishment of rural service centres, it does reflect on the way they are often used in regional development programmes.

Mobile services The concept of mobile services is a response to the high cost of establishing a rural service centre infrastructure, to the rigidity and inflexibility of the rural service centre concept, and to the difficult problem of the constraints of distance faced by a dispersed rural population. It is informed by the concept of periodic markets which have traditionally played a significant rôle in the rural areas of many developing countries (see Smith 1979, 1980). For example, in the Ankole region of Uganda, Good (1970) has documented a system of weekly produce markets and monthly and twice-monthly cattle markets which occur at different locations in the region each week or month. The mobile service concept is based upon the idea of taking services to people, as opposed to bringing people to services, which is the basis of rural service centres, and of using *time* in order to extend the effective range of a service. Essentially, it advocates that services, particularly marketing services, rotate over a large area in response to the 'need cycle' of recipients. The pattern of rotation is determined by the transportation network and services periodically locate close to major regional or district transportation routes, thereby maximizing accessibility. This tie with transportation sometimes leads to the simultaneous promotion of 'piggy-backing', for example, by attaching a mobile health service to rural school transportation.

Different services may have different rotational cycles and routes, but generally the agglomeration of functions to the greatest degree possible leads to a rationalization of trips. Significantly, it is expected that markets will specialize over time and that these periodic patterns will influence population distribution. Thus, a more permanent settlement system will develop over time, from the bottom up: some centres will consolidate and grow while others will disappear as the pattern of demand changes.

One difficulty with the concept is that it does not satisfactorily accommodate the locational requirements of constant services which cannot operate only once a week or once a month. This has led to the idea of tying periodic services and markets to a restricted rural centre concept. Periodic markets operate as one tier of the settlement hierarchy; the next tier up is at a considerably higher level (Reynolds 1981) in order to increase the range and level of services that are offered, albeit imperfectly, to rural dwellers. This does not accommodate the rationalization of basic services – primary education, health, potable water and so on, which require basic daily access – although theoretically it is possible for central places containing these activities to co-exist with periodic services and markets. Frequently, these basic services can be provided in villages, where these exist.

The 'holding centre' or 'key settlement' concept A cause of concern in many regions, particularly in developed countries, is the steady decline of small centres and thus the systematic erosion over time of the rural settlement system. Causes underpinning the decline are complex and vary with context, but four major causal factors can be identified. First, where a region declines economically, the younger and more educated people tend to migrate. Secondly, where changes in the structure of production occurs, such as the mechanization or increasing scale of agriculture, the number of people that can be supported in a region decreases. Thirdly, improvements in transportation and increases in private car ownership have resulted in many rural dwellers bypassing the smaller rural towns in favour of the shopping and service facilities of the larger centres. Fourthly, in scenically attractive areas, the purchase of holiday homes in small villages by wealthy inhabitants of the larger cities reduces the effective population of an area, as houses are only occupied for short periods of the year. Both of these factors can lead to a reduction in population thresholds. This results in the collapse of essential services: the grocer and butcher close down and per capita costs of village maintenance increase. Eventually, a point is reached where even those people who wish to stay are unable to (Cloke 1983).

The strategy most commonly advanced to combat this problem has been the promotion of holding towns or 'key villages': investment is concentrated in a few selected centres, which are encouraged to grow, in the belief that size itself is a protective factor. Those settlements that are not selected as key settlements are allowed to decline. In recent years, there has been a growing critique of this strategy and several alternatives to the problem have been postulated (see Cloke 1983). On one hand, advocates of a 'free market' ideology have continued to support the key settlement concept, but argue that rural services should only be provided at the locations in which they are most profitable. Unprofitable locations or settlements should receive even fewer resources than they have had in the past. On the other hand, there is growing support for a policy based on a greater dispersal of resources. A central criticism of the key settlement concept was that it did not allow sufficient resources for non-key settlements and a number of alternatives have been put forward (see Cloke 1983) in which either minimum functions are allocated to each village, or an arrangement is devised whereby essential services are shared between a set of small villages.

6 *Conclusion*

Settlement policy has become one of the main focuses of regional spatial planning in recent times, yet considerable confusion surrounds it. In particular, its rôle in, and the form of its interpenetration with, broader approaches to development planning, are frequently misunderstood. This book has attempted to identify the main features of the major approaches to regional development which have been promoted over the last few decades, to explore the implications of these for settlement policy, particularly in developing countries, and to clarify the central debates relating to the settlement system and its rôle in national and regional development. The review leads to two major interrelated observations about developments within the field of settlement policy.

The first is that although shifts in the theory and practice of development logically demanded certain adjustments in the form of settlement policy, there are several basic ideas about settlements and their rôle in development which have remained remarkably consistent. In reviewing changes in thinking about settlement policy in the postwar period, therefore, it would be more accurate to refer to *shifts in emphasis* rather than fundamental change. One of the most pervasive of these consistencies has been the belief, common to the various approaches to development, in the positive rôle played by an integrated settlement hierarchy. Settlement debates, rather than questioning the significance of hierarchical integration *per se*, have concentrated more on the issue of the *level* of the hierarchy that should be stimulated and whether developmental impulses flow downwards from the top of the hierarchy or upwards from the region and the small towns at the base of the hierarchy. Until the 1970s, the dominant emphasis in development programmes was on the concentration of development efforts at the top end of the hierarchy. Since that time there has been a shift in emphasis towards the lower levels of the hierarchy, to 'development from below' and to a view of settlements as simply supportive in the development process rather than as the central and dynamic agents of development. Related to this have been debates about appropriate size and spatial distributions of settlements and the various tools which should be used to stimulate different levels of the hierarchy.

The second observation, related to the first, is that settlement policy, both in theory and in practice, has tended to operate relatively autonomously from development theory. While settlement planning has responded to paradigmatic shifts in the theory and practice of development to some degree, it has nonetheless retained a concern with specifically *spatial* issues.

These two tendencies which have characterized spatial settlement planning have, to some extent, been responsible for a number of problems that have arisen in this policy arena. First, there has been a tendency to view settlement issues as problems in their own right, which can be dealt with exclusively through the mechanisms of spatial planning. One reason for this is the prevalence of a 'spatial separatist theme' in regional theory, that is, the 'notion that it is possible to identify, separate and evaluate the *spatial* as an independent phenomenon or a property of events examined through spatial analysis' (Sack 1974, p. 1). Regional theorists have tended to treat space as separate from social processes, hence spatial patterns in themselves are seen as problems that require intervention. Thus, for example, the lack of a settlement hierarchy, the existence of a primate city system, or the absence of small towns, are seen as problems in themselves. As a result, social and economic issues are often either ignored or packed uncomfortably into the definition of 'spatial problems'. The danger in this is that spatially uneven development and an atypical settlement hierarchy may in fact be and usually are the result of economic and political forces which lie beyond the scope of settlement policy. For example, the failure of growth poles to bring about a change in the pattern of development in a country may result as much from the absence of decentralizing forces within the space economy as from problems associated with the implementation of growth pole theory itself.

Secondly, the relative autonomy of the settlement question together with the emphasis on settlement issues over time has resulted in settlement policy taking on the form of standardized approaches, regardless of the vagaries of context. In this process, there has also been a tendency to attempt to solve very different problems using the same settlement policy package. In these cases, tools that are at best appropriate to only a part of the problem addressed are used in an uncritical way. Significantly, this issue of the adoption of standardized, generalized approaches extends beyond settlement policy. Although there is undoubtedly some flexibility of interpretation regarding the approaches to development of the kind reviewed here, there is nevertheless a tendency for some commentators to argue for the *inherent*, universal superiority of certain approaches to initiating processes of development over others, regardless of contextual differences. In part, this results from the way in which these approaches have been formulated. In part, too, it results from the uncritical way in which development theory has been adopted and used by decision makers. Whatever the cause, the negative consequences are likely to be severe.

Thirdly, since settlement planning is seen as an area of intervention in its own right, too much is expected of it: it is often expected to have consequences which it is unable to generate. For example, a system of service centres may provide access to a range of services – provided it is adequately planned – but it will not necessarily spark economic development, since other factors outside the realm of settlement planning *per se* may well prevent this. Similarly, growth poles might provide a limited number of jobs in peripheral regions, but they are unlikely to diffuse development to a significant degree, or radically change patterns of spatial inequality.

Fourthly, a relatively unquestioned focus upon spatial issues may be reflected in the goals of development programmes and thereby limit the extent to which the fundamental determinants of development are addressed. Gore (1984) makes the point that spatial policies can have only the limited effect of altering the spatial distribution of growth or welfare. The reason for this is that 'spatial policies are innately conservative in the sense that they do not seek to affect the underlying processes of social and economic change' (Gore 1984, p. 221). It is not surprising, therefore, that spatial policies have had such limited success, particularly if success is measured in terms of factors reflecting human conditions such as interpersonal income convergence or reductions in poverty rather than – as is the case generally – of spatial conditions such as greater interregional convergence of income and welfare. It is significant that these types of convergence are not necessarily best promoted through the same types of policies or programmes: it is quite possible to promote one without the other. The point is not of academic interest only. As Gilbert and Gugler (1982) point out, policies to promote greater interregional convergence are generally more compatible with growth-orientated approaches to development than interpersonal, redistributionary policies.

At a general level, these trends and tendencies provide a number of basic lessons for settlement planning.

The first is the importance of *context* in determining how the development problem is defined and the nature of appropriate policies or programmes. This recognition necessitates a planning approach which starts from the particular development problem at hand, which understands the social and economic processes underlying it and which formulates clear and specific goals and strategies in relation to the problem. The decision as to whether, or to what extent, interference in the settlement system is justified should result from this process: it hinges on the extent to which spatial and settlement issues are contributory factors in the development problem. The inclusion of settlement planning in the final policy package would then occur in response to an understanding of context and as an adjunct to broader policy issues and not as a solution in its own right. Finally, the particular tools that are used in relation to settlement planning should also be determined by an understanding of the development problem at hand, and not simply employed as part of a standardized, pre-packaged settlement 'model'. An important part of this broader approach to developmental issues is to avoid thinking in mutually exclusive terms about dimensions of the development problem. One arena where this is currently particularly apparent in development planning is the dichotomization between 'top down' and 'bottom up' approaches. In most development programmes, both 'top down' decisions and actions – defined as those initiated from a centralized source beyond the control of local communities – and 'bottom up' decisions and actions – those under the control of local communities – are required: they have the potential to achieve very different developmental objectives. Further, they are interrelated: a framework of 'top down' actions often creates the conditions necessary for the successful implementation of 'bottom up' programmes, but 'bottom up' actions are usually the ones that have the greatest potential to affect positively the lives of the poor. However, there are often *inherent* contradictions and conflicts between the two that are irreducible. The best that can be achieved in these situations is to recognize the conflicts and apply judgement in the design of policies and programmes to minimize the disruptive effects of these. A failure to accept the reality of conflict in development planning, however, often leads to the exclusive advocacy of one 'approach' at the expense of the others.

The second lesson, following on from the first, is that standardized settlement approaches may be useful, but this cannot be assumed *a priori*. The nature of the development problem may in fact require that *elements* of the various settlement packages are used in combination to bring about the desired result. Of significance here is an understanding of the assumptions which underlie the various approaches to settlement planning and the extent to which these assumptions hold true in the context being considered.

The third lesson is that settlement planners must recognize the limits of their expertise. To believe that underdevelopment can be effected significantly through the manipulation of the spatial pattern of settlement alone is to fall into the trap of spatial determinacy. There are two aspects to this. On one hand, the complexity of most development problems is such that a full, integrated policy package will be required: one which tackles economic, social, political, ideological and cultural issues, and which may have both spatial and aspatial aspects to it. The often limited rôle of settlement planning in these packages must be recognized and defined. On the other hand, the obstacles to this rôle being fully realized in the particular context must also be identified. For example, it cannot be assumed that the state will play a neutral rôle in the development process: in certain circumstances, it may well be directly obstructive. Further, development policies may be hindered or obstructed by the particular form of the class structure, by the nature of ruling class alliances or by the form and nature of capitalist development. The issue of what is possible at any point in time must therefore be faced squarely.

The fourth lesson is that all planning, including settlement planning, is a highly political process: planners cannot perceive themselves as neutral agents in the development process. Different approaches to development and settlement planning will have different outcomes for various groups in society. For example, the redistributionary approaches of the mid-1970s have greater potential to promote the interests of the poorest groups and to promote equity than accelerated growth approaches which have been characterized as an attack on the interests of the working classes. None the less, the ability of planners to influence the course of development policy will depend, ultimately, on the nature of the ruling class, the strength of the other classes and on economic and political conditions. In the final analysis, as Gilbert and Gugler (1982, p. 197) conclude, developmentally positive advances 'are not made by technicians on the basis of objective studies. Rather they stem from political decisions made ultimately by the majority of the population'.

References

Alden, J. and R. Morgan 1974. *Regional planning: a comprehensive view*. Leighton Buzzard, UK: Leonard Hill.

Allison, C. and R. Green 1983. Stagnation and decay in sub-Saharan Africa: dialogues, dialectics and doubts. *Institute of Development Studies Bulletin* **14**(1), 1–10.

Alonso, W. 1972. The question of city size and national policy. In *Recent developments in regional science*, R. Funck (ed.), 111–18. London: Pion.

Alonso, W. 1975. Urban and regional imbalances in economic development. In *Regional policy: readings in theory and applications*, J. Friedmann and W. Alonso (eds), 622–35. Cambridge, Mass.: MIT Press.

Anderson, J. 1978. *The political economy of urbanism: an introduction and bibliography*. London: Department of Urban and Regional Planning, Architectural Association.

Babarovic, I. 1978. Rural marginality and regional development policies in Brazil. In *Regional policies in Nigeria, India and Brazil*, A. Kuklinski (ed.), 194–264. The Hague: Mouton.

Balassa, B. 1984. Adjustment policies in developing countries: a reassessment. *World Development* **12**(9), 955–72.

Baldwin, R. E. 1956. Patterns of development in newly settled regions. *Manchester School of Economic and Social Studies* **24**(2), 161–79.

Baster, N. 1973. Development indicators: an introduction. *Journal of Development Studies* **8**(3), 1–20.

Beng, C. H. 1980. Export-oriented industrialisation and dependent development: the experience of Singapore. *Institute of Development Studies Bulletin* **12**(1), 35–41.

Berg, A. 1980. A strategy to reduce malnutrition. In *Poverty and basic needs*, World Bank (ed.), 18–22. Washington: World Bank.

Berger, P. L. 1974. *Pyramids of sacrifice*. London: Penguin.

Berry, B. J. L. 1961. City size distributions and economic development. *Economic Development and Cultural Change* **9**(4), 573–88.

Berry, B. J. L. 1972. Hierarchical diffusion: the basis of developmental filtering and spread in a system of growth centres. In *Growth centres in regional economic development*, N. M. Hansen (ed.), 108–38. New York: Free Press.

Berry, B. J. L. and W. L. Garrison 1958. Recent developments in central place theory. *Papers and Proceedings of the Regional Science Association* **4**, 107–21.

Boudeville, J. R. 1969. *Problems of regional economic planning*. Edinburgh: Edinburgh University Press.

Brookfield, H. 1975. *Interdependent development*. London: Methuen.

Burki, S. J. 1980. Sectoral policies for meeting basic needs. In *Poverty and basic needs*, World Bank (ed.), 13–19. Washington, DC: World Bank.

Burki, S. J. and M. Ul Haq 1981. Meeting basic needs: an overview. *World Development* **9**, 13–17.

Carapetis, S., H. L. Beenhakker and J. D. F. Howe 1984. *The supply and quality of rural transport services in developing countries: a comparative review*. World Bank Staff Working Paper 654.

Chenery, M., M. S. Ahluwalia, C. L. G. Bell, O. Duloy and R. Jolly 1974. *Redistribution with growth*. London: Oxford University Press.

Christaller, W. 1966. *The central places of southern Germany* (translated by C. W. Baskin). Englewood Cliffs, N.J.: Prentice-Hall.

Christiansen, R. and J. Kydd 1983. *An economic assessment of the rural centres programme*. Unpublished report to Malawi Government.

Cleveland, H. van B. and W. H. Brittain 1978/79. Are the L.D.C.'s in over their heads? *Economic Impact* **22**, 49–53.

Cline, W. R. 1982. Can the East Asian model of development be generalized? *World Development* **10**(2), 81–90.

Cloke, P. J. 1983. *An introduction to rural settlement planning*. London: Methuen.

Colclough, C. 1983. Are African governments as unproductive as the accelerated development report implies? *Institute of Development Studies Bulletin* **14**(1), 24–9.

Dahrendorf, R. 1968. *Essays in the theory of society*. California: Stanford University Press.

Darwent, D. F. 1969. Growth poles and growth centres in regional planning – a review. *Environment and Planning* **1**, 5–32.

Dewar, D., A. Todes and V. Watson 1982. *Urbanisation: responses and policies in four case studies – Kenya, Zambia, Tanzania and Zimbabwe*. University of Cape Town, UPRU Working Paper 25.

Dewar, D., A. Todes and V. Watson 1984. *Issues of regional development in peripheral regions of South Africa, with particular reference to settlement policy: the case of the Transkei*. University of Cape Town, UPRU Working Paper 29.

Dewar, D., A. Todes and V. Watson 1985. *A framework for national settlement strategy in South Africa*. Final report of the Council for Scientific and Industrial Research, Pretoria.

Dos Santos, T. 1973. The crisis in development theory and the problem of development in Latin America. In *Underdevelopment and development. The Third World today*, H. Bernstein (ed.), 57–80. London: Penguin.

Douglass, M. 1981. Thailand: territorial dissolution and alternative regional development for the central plains. In *Development from above or below? The dialectics of regional planning in developing countries*, W. B. Stöhr and D. R. F. Taylor (eds), 183–207. New York: Wiley.

Dumont, R. 1979. *Towards another development in rural Zambia*. Report of the Institut Nationale Agronomique de Paris.

Duncan, O. D. 1957. Optimum sizes of cities. In *Cities and society: the revised reader in urban sociology*, P. K. Hall and A. J. Reiss Jnr (eds), 759–72. Glencoe: Free Press.

Dunford, M., M. Geddes and D. Perrons 1981. Regional policy and the crisis in the U.K.: a long-run perspective. *International Journal of Urban and Regional Research* **5**(3), 377–409.

Dwyer, D. J. 1977. Economic development: development for whom? *Geography* **62**, 325–44.

Edel, C. K., M. Edel, K. Fox, A. Markusen, P. Meyer and D. Vail 1978. Uneven regional development: an introduction to this issue. *Review of Radical Political Economy* **10**(3), 1–11.

El Shakhs, S. S. M. 1965. *Development primacy and the structure of cities*. Ph.D dissertation, Harvard University.

El Shakhs, S. S. M. 1974. Development planning in Africa: an introduction. In *Urbanization, national development and regional planning*, S. S. M. El Shakhs and R. Obudho (eds), 3–12. New York: Praeger.

Evans, D. and P. Alizadeh 1984. Trade, industrialisation and the visible hand. *Journal of Development Studies* **21**(1), 22–46.

Fair, T. J. D. 1982. *South Africa: spatial frameworks for development*. Kenwyn: Juta.

Fishlow, A. 1984. Summary comment on Adelman, Balassa and Streeten. *World Development* **12**(9), 979–82.

Frank, A. G. 1967. *Capitalism and underdevelopment in Latin America: historical studies of Chile and Brazil*. New York: Monthly Review Press.

Frank, A. G. 1971. *Capitalism and underdevelopment in Latin America*. London: Penguin.

Friedmann, J. 1966. *Regional development policy: a case study of Venezuela*. Cambridge, Mass.: MIT Press.

Friedmann, J. 1968. The strategy of deliberate urbanisation. *Journal of the American Institute of Planners* **34**(6), 364–73.

Friedmann, J. 1972. A general theory of polarised development. In *Growth centres in regional economic development*, N. M. Hansen (ed.), 82–107. New York: Free Press.

Friedmann, J. 1973. *Urbanisation, planning and national development*. Beverley Hills: Sage.

Friedmann, J. 1974. *A spatial framework for rural development. Problems of organization and implementation*. Report of the US Agency for International Development.

Friedmann, J. 1980. The territorial approach to rural development in the People's Republic of Mozambique: six discussion papers. *International Journal of Urban and Regional Research* **4**(1), 97–115.

Friedmann, J. 1981/82. Regional planning for rural mobilization in Africa. *Rural Africana* **12–13**, 3–19.

Friedmann, J. 1982. *Political and technical movements in development: agropolitan development revisited.* Paper presented at the Commission for International Geography Union, Minas Gerais, Brazil.

Friedmann, J. and W. Alonso (eds) 1975. *Regional policy: readings in theory and applications.* Cambridge, Mass.: MIT Press.

Friedmann, J. and C. Weaver 1979. *Territory and function.* London: Edward Arnold.

Fuchs, V. 1967. *Differentials in hourly earnings by region and city size 1959.* National Bureau for Economic Research, Occasional Paper 101. New York: Columbia University Press.

Funnell, D. C. 1973. Rural business centres in a low income economy. Some theoretical problems. *Tijdschrift voor Economiese en Sociale Geografie* **64**(2), 86–92.

Ghai, D. P. 1980. What is a basic needs approach to development all about? In *Basic needs – approaches to development: some issues regarding concepts and methodology*, D. P. Ghai, A. R. Khan, E. L. H. Lee and T. Alfthon (eds), 19–59. Geneva: ILO.

Gilbert, A. G. 1975. A note on the incidence of development in the vicinity of a growth centre. *Regional Studies* **9**, 325–33.

Gilbert, A. 1976. The argument for very large cities reconsidered. *Urban Studies* **13**(1), 27–34.

Gilbert, A. 1977. The argument for very large cities reconsidered. A reply. *Urban Studies* **14**(2), 225–8.

Gilbert, A. and J. Gugler 1982. *Cities, poverty and development: urbanization in the Third World.* Oxford: Oxford University Press.

Godfrey, M. 1983. Export orientation and structural adjustment in sub-Saharan Africa. *Institute of Development Studies Bulletin* **14**(1), 39–44.

Goldsmith, W. W. 1982. Enterprise zones: if they work we're in trouble. *International Journal of Urban and Regional Research* **6**(3), 435–42.

Good, M. 1970. *Rural markets and trade in East Africa.* University of Chicago Department of Geography Research Paper 128.

Gore, C. 1984. *Regions in question. Space, development theory and regional policy.* London: Methuen.

Gottschalk, K. 1977. Industrial decentralization, jobs and wages. *South African Labour Bulletin* **3**(5), 50–8.

Goulet, D. 1979. Development as liberation: policy lessons from case studies. *World Development* **7**(6) 555–66.

Green, R. 1978. Basic human needs: concept or slogan, synthesis or smokescreen? *Institute of Development Studies Bulletin* **9**(4), 7–11.

Green, R. 1983. Incentives, policies, participation and response: reflections on World Bank policies and priorities in agriculture. *Institute of Development Studies Bulletin* **14**(1), 30–8.

Griffiths-Jones, S. 1983. A Chilean perspective. *Institute of Development Studies Bulletin* **14**(1), 50–4.

Hägerstrand, T. 1953. *Innovation–diffusion as a spatial process* (translated 1967). Chicago: Chicago University Press.

Hall, P. 1982. Enterprise zones: a justification. *International Journal of Urban and Regional Research* **6**(3), 416–21.

Hansen, N. M. 1981. Development from above: the centre-down development paradigm. In *Development from above or below? The dialectics of regional planning*, W. B. Stöhr and D. R. F. Taylor (eds), 15–38. New York: Wiley.

Harrison, B. 1982. The political and economic origins of the urban enterprise zone proposal: a critique. *International Journal of Urban and Regional Research* **5**(3), 422–8.

Harrison, J. 1978. *Marxist economics for socialists. A critique of reformism.* London: Pluto Press.

Harvey, D. 1973. *Social justice and the city*. London: Edward Arnold.

Henn, J. K. 1983. Feeding the cities and feeding the peasants. What role for Africa's women farmers? *World Development* **11**(12), 1043–55.

Hermansen, T. 1972a. Development poles and development centres in national and regional development. Elements of a theoretical framework. In *Growth poles and growth centres in regional planning*, Vol. 5, A. Kuklinski (ed.), 1–67. The Hague: Mouton.

Hermansen, T. 1972b. Development poles and related theories: a synoptic review. In *Growth centres in regional economic development*, N. M. Hansen (ed.), 160–203. New York: Free Press.

Hicks, N. L. 1980. Is there a trade-off between growth and basic needs? In *Poverty and basic needs*, World Bank (ed.), 23–5. Washington, DC: World Bank.

Higgens, B. 1972. Growth pole policy in Canada. In *Growth centres in regional economic development*, N. M. Hansen (ed.), 266–81. New York: Free Press.

Hirschman, A. O. 1958. *The strategy of economic development.* New Haven, Conn.: Yale University Press.

Hoch, I. 1972. Income and city size. *Urban Studies* **9**(3), 299–328.

Holland, S. 1971. Regional underdevelopment in a developed economy: the Italian case. *Regional Studies* **5**, 71–90.

Hyden, G. 1980. *Beyond Ujamaa in Tanzania. Underdevelopment and an uncaptured peasantry*. Nairobi: Heinemann.

Illich, I. 1971. *Deschooling society*. London: Calder & Boyars.

Innis, H. A. 1930. *The fur trade in Canada: an introduction to Canadian economic history*. Toronto: Toronto University Press.

International Labour Office 1972. *Employment, incomes and equality. A strategy for increasing productive employment in Kenya.* Geneva: ILO.

International Labour Office 1976. *Employment, growth and basic needs. A one world problem*. Geneva: ILO.

Isard, W. 1956. *Location and space economy*. New York: Wiley.

Isard, W. 1960. *Methods of regional analysis: an introduction to regional science.* New York: Technology Press Massachusetts. Institute of Technology and Wiley.

Johnson, E. A. J. 1970. *The organization of space in developing countries.* Cambridge, Mass.: Harvard University Press.

Jolly, R. 1977., Changing views on development. In *Surveys for development*, J. J. Nossin (ed.), 19–36. Amsterdam: Elsevier.

Jones, P. J. 1983. France's population disperses. *Town and Country Planning* **52**(3), 81–3.

Jones, T. 1984. Entrepreneurial talent and the call for deregulation. A critical response to the Swart Commission's proposals for economic development in the Ciskei. *Indicator South Africa. Rural Monitor* **2**(2), 8–10.

Kaplinsky, R. 1984. The international context for industrialisation in the coming decade. *Journal of Development Studies* **21**(1), 75–96.

Klaassen, L. H. 1965. *Area development policies in Britain and the countries of the Common Market*. Washington: US Department of Commerce.

Kuznets, S. S. 1966. *Growth: rate, structure and spread*. New Haven: Yale University Press.

Laclau, E. 1971. Feudalism and capitalism in Latin America. *New Left Review* **67**, 19–38.

Lasuén, J. R. 1972. On growth poles. In *Growth centres in regional economic development*, N. M. Hansen (ed.), 20–49. New York: Free Press.

Lefeber, L. and M. Datta-Chaudhuri 1971. *Regional development experiences and prospects in south and south-east Asia*. The Hague: Mouton.

Lele, U. 1975. *The design of rural development. Lessons from Africa*. Baltimore: Johns Hopkins University Press.

Lewis, J. P. 1962. *Quiet crisis in India: economic development and American policy*. Washington: Brookings Institution.

Leys, C. 1975. *Underdevelopment in Kenya*. London: Heinemann.

Lienbach, T. R. 1983. Transport evaluation in rural development. An Indonesian case study. *Third World Planning Review* **5**(1), 21–35.

Lipton, M. 1977. *Why poor people stay poor: a study of urban bias in world development*. London: Temple Smith.

Lloyd, P. E. and P. Dicken 1979. *Location in space*, 3rd edn. New York: Harper & Row.

Lösch, A. 1954. *The economics of location*. Translated by W. Woglam and W. F. Stolper. New Haven, Conn.: Yale University Press.

Mabogunje, A. L. 1978. Growth poles and growth centres in the regional development of Nigeria. In *Regional policies in Nigeria, India and Brazil*, A. Kuklinski (ed.), 8–104. The Hague: Mouton.

Mabogunje, A. L. 1980. *The development process: a spatial perspective*. London: Hutchinson University Library for Africa.

McCall, M. K. 1977. Political economy and rural transport: a reappraisal of transportation impacts. *Antipode* **9**, 56–67.

Mascarenhas, A. C. and C. F. Claeson 1972. Factors influencing Tanzania's urban policy. *African Urban Notes* **6**(3), 24–41.

Massey, D. 1978 Regionalism: some current issues. *Capital and Class* **6**, 106–25.

Massey, D. 1981. The U.K. electrical engineering and electronics industries: the implications of the crisis for the restructuring of capital and locational change. In *Urbanisation and urban planning in capitalist society*, M. Dear and A. J. Scott (eds), 199–230. New York: McGraw-Hill.

Massey, D. 1982. Enterprise zones: a political issue. *International Journal of Urban and Regional Research* **6**(3), 429–34.

Mayer, J. 1979. Spatial aspects of basic needs strategy: the distribution of essential services. *International Labour Review* **8**(1), 59–74.

Meadows, D. H., D. L. Meadows, J. Ronders and W. W. Behrens III 1972. *The limits to growth. A report to the Club of Rome*. New York: Universe Books.

Mera, K. 1973. On urban agglomeration and economic efficiency. *Economic Development and Cultural Change* **21**(2), 309–24.

Mesavoric, M. and E. Pestel 1974. *Mankind at the turning point. The second report to the Club of Rome*. New York: Dutton.

Mills, E. S. 1972. Welfare aspects of national policy towards city sizes. *Urban Studies* **19**(1), 117–28.

Misra, P. S. and K. V. Sundaram 1978. Growth foci as instruments of modernization in India. In *Regional policies in Nigeria, India and Brazil*, A. Kuklinski (ed.), 105–93. The Hague: Mouton.

Mlia, N. 1975. National urban development policy: the issues and options. In *Urbanisation, national development and regional planning in Africa*, S. El Shakhs and R. Obudho (eds), 75–89. New York: Praeger.

Moseley, M. J. 1973a. The impact of growth centres in rural regions, I. An analysis of spatial 'patterns' in Brittany. *Regional Studies* **7**(1), 57–75.

Moseley, M. J. 1973b. The impact of growth centres in rural regions, II. An analysis of spatial 'flows' in East Anglia. *Regional Studies* **7**(1), 77–94.

Moseley, P. 1978. Implicit models and policy recommendations: policy towards the 'informal sector' in Kenya. *Institute of Development Studies Bulletin* **9**(3), 3–10.

Moser, C. O. 1978. Informal sector or petty commodity production: dualism and dependence in urban development. *World Development* **6**(9/10), 1041–64.

Myrdal, G. 1957. *Economic theory and underdeveloped regions*. London: Duckworth.

Nattrass, J. 1982. *The research requirements of the basic needs approach to development in the South African context*. Draft paper presented at the Bophuthatswana Rural Development Seminar, University of Bophuthatswana.

North, P. C. 1955. Location theory and regional economic growth. *Journal of Political Economy* **LXIII**, 243–58.

O'Connor, A. M. 1976. Third World or one world? *Area* **8**, 269–71.

Parr, J. B. 1973. Growth poles, regional development and central place theory. *Papers of the Regional Science Association* **31**, 173–212.

Pedersen, P. O. 1975. *Urban–regional development in South America: a process of diffusion and integration*. The Hague: Mouton.

Perkins, D. 1978. Meeting basic needs in the People's Republic of China. *World Development* **6**(5), 561–6.

Perloff, H. S., E. S. Dunn Jnr, E. E. Lampard and R. F. Muth 1960. *Regions, resources and economic growth*. Baltimore: Johns Hopkins University Press.

Perroux, F. 1955. Note on the concept of growth poles. *Economie Appliquée* **8**, 307–20.

Perry, D. C. and A. J. Watkins 1981. Contemporary dimensions of uneven urban development. In *City, capital and class*, M. Harloe and E. Lebas (eds), 115–42. London: Edward Arnold.

Phillips, A. 1977. The concept of development. *Review of African Political Economy* **8**, 7–20.

Piöro, Z. 1972. Growth poles and growth centres theory as applied to settlement development in Tanzania. In *Growth poles and growth centres in regional planning*, Vol. 5, A. Kuklinski (ed.), 169–94. The Hague: Mouton.

Pottier, O. 1963. Axes de communication et développement économique. *Revue Economique* **14**, 113–14.

Prantilla, E. B. 1979. Regional development and planning: the Philippine experience. In *Growth pole strategy, regional development and planning in Asia*, UNCRD (ed.), 83–101. Nagoya: UNCRD.

Preston-Whyte, E. and S. Nene 1984. *Where the informal sector is not the answer: women and poverty in KwaZulu*. Paper presented at the Second Carnegie Conference, Cape Town, South Africa.

Raikes, P. L. 1975. Ujaama and rural socialism. *Review of African Political Economy* **3**, 33–52.

Renaud, B. 1981. *National urbanisation policy in developing countries*. Published for the World Bank by Oxford University Press.

Reynolds, N. E. 1981. *The utility of a combined periodic service and regulated market system in the development of Zimbabwe's tribal trust lands*. Paper presented at Zimbabwe Economic and Social Conference on Rural Development.

Richardson, H. W. 1972. Optimality in city size, systems of cities and urban policy: a sceptic's view. *Urban Studies* **9**, 29–48.

Richardson, H. W. 1976. The argument for very large cities reconsidered. A comment. *Urban Studies* **13**, 307–10.

Richardson, H. W. 1981. National urban development strategies in developing countries. *Urban Studies* **18**, 267–83.

Riddell, R. C. 1977. *Alternatives to poverty*. Gwelo: Mambo Occasional Papers Socio-Economics. Series, no. 1.

Riddell, R. C. 1978. *The land problem in Rhodesia – alternatives for the future*. Gwelo: Mambo Occasional Papers Socio-Economics. Series, no. 11.

Robinson, G. and K. B. Salih 1971. The spread of development around

Kuala Lumpur: a methodology for an exploratory test of some assumptions of the growth pole model. *Regional Studies* **5**(4), 305–14.
Rondinelli, D. A. 1983. *Secondary cities in developing countries*. Beverley Hills, Calif.: Sage.
Rondinelli, D. A. and K. Ruddle 1976. *Urban functions in rural development: an analysis of integrated spatial development policy*. Report of the Office of Urban Development Technical Assistance Bureau, Agency for International Development, US Department of State.
Rondinelli, D. A. and K. Ruddle 1978. *Urbanisation and rural development. A spatial policy for equitable growth*. New York: Praeger.
Rostow, W. W. 1960. *The stages of economic growth – a non-Communist manifesto*. Cambridge: Cambridge University Press.

Sack, R. D. 1974. The spatial separatist theme in geography. *Economic Geography* **50**, 1–9.
Salau, A. T. 1979. Urbanization, planning and public policies in Nigeria. In *Development of urban systems in Africa*, R. A. Obudho and S. El Shakhs (eds), 196–207. New York: Praeger.
Sandbrook, R. 1982. *The politics of basic needs. Urban aspects of assaulting poverty in Africa*. London: Heinemann.
Sayer, A. 1981. Review of Friedmann, J. and C. Weaver, 1979. Territory and function: the evolution of regional planning; London: Edward Arnold. *International Journal of Urban and Regional Research* **5**(4), 596–8.
Schmitz, H. 1984. Industrial strategies in less developed countries: some lessons of historical experience. *Journal of Development Studies* **21**(1), 1–21.
Schumacher, E. F. 1973. *Small is beautiful: a study of economics as if people mattered*. London: Blond & Briggs.
Seers, D. 1972. What are we trying to measure? In *Measuring development*, N. Baster (ed.), 21–36. London: Cass.
Shindman, B. 1955. An optimum size for cities. *Canadian Geographer* **5**, 85–8.
Slater, D. 1974. Contribution to a critique of development geography. *Revue Canadienne des Etudes Africaines* **8**(2), 325–54.
Smith, R. H. T. 1979. Periodic market-places and periodic marketing: review and prospect – I. *Progress in Human Geography* **3**(4), 471–505.
Smith, R. H. T. 1980. Periodic market-places and periodic marketing: review and prospect – II. *Progress in Human Geography* **4**(1), 1–31.
Soja, E. W. 1976. *The geography of modernization: a radical reappraisal*. Special lecture, School of Oriental and African Studies, London. Unpublished.
Soja, E. W. 1982. *Spatiality, politics and the role of the state*. Paper presented at the Latin American Congress of International Geography Union, Rio de Janeiro.
Soja, E. W. and C. E. Weaver 1976. Urbanisation and underdevelopment in East Africa. In *Urbanisation and counter-urbanisation*, B. J. L. Berry (ed.), 233–66, Beverley Hills: Sage.
Stewart, F. 1980. Country experience in providing for basic needs. In

Poverty and basic needs, World Bank (ed.), 9–12. Washington, DC: World Bank.

Stöhr, W. B. 1974. *Inter-urban systems and regional economic development*. Association of American Geographers Resource Paper no. 26.

Stöhr, W. B. 1975. *Regional development: experiences and prospects in Latin America*. The Hague: Mouton.

Stöhr, W. B. 1980. *Development from below: the bottom-up and periphery-inward paradigm*. Interdisciplinary Institute of Urban and Regional Studies, Discussion 6. Vienna: University of Economics.

Stöhr, W. B. 1981. Development from below: the bottom-up and periphery-inward paradigm. In *Development from above or below? The dialectics of regional planning in developing countries*, W. B. Stöhr and D. R. F. Taylor (eds), 39–69. New York: Wiley.

Stöhr, W. B. and D. R. F. Taylor (eds) 1981. *Development from above or below? The dialectics of regional planning in developing countries*. New York: Wiley.

Stöhr, W. B. and F. Tödtling 1977. Spatial equity – some anti-theses to current regional development doctrine. *Papers of the Regional Science Association* **38**, 33–53.

Stöhr, W. B. and F. Tödtling 1978. An evaluation of regional policies – experiences in market and mixed economies. In *Human settlement systems: international perspectives on structure, change and public policy*, N. M. Hansen (ed.), 85–119. Cambridge, Mass.: Balinger.

Streeten, P. 1980. From growth to basic needs. In *Poverty and basic needs*, World Bank (ed.), 5–8. Washington: World Bank.

Streeten, P. 1981. *Development perspectives*. London: Macmillan.

Streeten, P. 1982a. Approaches to a new international economic order. *World Development* **10**(1), 1–17.

Streeten, P. 1982b. Development ideas in historical perspective. *Economic Impact* **40**, 11–19.

Streeten, P.and S. J. Burki 1978. Basic needs: some issues. *World Development* **6**(3), 411–21.

Sundaram, K. V. 1980. Comment. *Regional Development Dialogue* **1**, 98–101.

Thomas, M. D. 1972. Growth pole theory: an examination of some of its basic concepts. In *Growth centres in regional economic development*, N. M. Hansen (ed.), 50–81. New York: Free Press.

Todd, D. 1974. An appraisal of the development pole concept in regional analysis. *Environment and Planning A* **6**, 291–306.

Tokman, V, 1978. An exploration into the nature of informal–formal sector relationships. *World Development* **6**(9/10), 1065–75.

Tuppen, J. N. 1983. The development of French new towns: an assessment of progress. *Urban Studies* **20**, 11–30.

Turner, J. F. C. 1976. *Housing by people. Towards autonomy in building environments*. London: Marion Boyars.

Ul Haq, M. 1981. An international perspective on basic needs. *Development Studies of Southern Africa* **4**(1), 135–41.

Uphoff, N. R. and M. J. Esmain 1974. *Local organization for rural development: analysis of Asian experience*. Rural Development Committee, Centre for International Studies, Cornell University.

Von Freyhold 1979. *Ujamaa villages in Tanzania. Analysis of a social experiment*. Nairobi: Heinemann.

Warren, B. 1972. Capitalist planning and the state. *New Left Review* **72**, 3–29.

Weaver, C. 1981. Development theory and the regional question. A critique of spatial planning and its detractors. In *Development from above or below? The dialectics of regional planning in developing countries*, W. B. Stöhr and D. R. F. Taylor (eds), 73–98. New York: Wiley.

Wells, J. 1981. Monetarism in the U.K. and the southern Cape: an overview. *Institute of Development Studies Bulletin* **13**(1), 14–25.

Whitsun Foundation 1980. *Rural service centres development study*. Salisbury: Whitsun Foundation.

Willbanks, T. J. 1972. Accessibility and technological change in northern India. *Annals of the Association of American Geographers* **62**(3), 427–36.

Wilson, G. W., B. R. Bergman, L. V. Hirsch and M. S. Klein 1966. *The impact of highway investment on development*. Washington: Brookings Institution Transport Research Programme.

Wingo, L. Jnr 1972. Issues in a national urbanisation policy for the United States. *Urban Studies* **9**, 3–28.

Wolf-Phillips, L. 1979. Why Third World? *Third World Quarterly* **1**, 105–16.

World Bank 1981. *Accelerated development in sub-Saharan Africa. An agenda for action*. Washington, DC: World Bank.

Yawitch, J. 1982. *Betterment: the myth of homeland agriculture*. Johannesburg: South African Institute of Race Relations.

Zingel, J. 1984. Land tenure modernization and rural development in the Ciskei. A consideration of some basic issues raised by the Swart Commission approach. *Indicator South Africa. Rural Monitor* **2**(2), 3–7.

Zipf, G. K. 1949. *Human behaviour and the principle of least effort*. Cambridge: Cambridge University Press.

Index